T0237306

SpringerBriefs in Molecular Science

Chemistry of Foods

Series Editor

Salvatore Parisi, Lourdes Matha Institute of Hotel Management and Catering
Technology, Kerala, India

The series Springer Briefs in Molecular Science: Chemistry of Foods presents compact topical volumes in the area of food chemistry. The series has a clear focus on the chemistry and chemical aspects of foods, topics such as the physics or biology of foods are not part of its scope. The Briefs volumes in the series aim at presenting chemical background information or an introduction and clear-cut overview on the chemistry related to specific topics in this area. Typical topics thus include:

- Compound classes in foods—their chemistry and properties with respect to the foods (e.g. sugars, proteins, fats, minerals, …)
- Contaminants and additives in foods—their chemistry and chemical transformations
- Chemical analysis and monitoring of foods
- Chemical transformations in foods, evolution and alterations of chemicals in foods, interactions between food and its packaging materials, chemical aspects of the food production processes
- Chemistry and the food industry—from safety protocols to modern food production

The treated subjects will particularly appeal to professionals and researchers concerned with food chemistry. Many volume topics address professionals and current problems in the food industry, but will also be interesting for readers generally concerned with the chemistry of foods. With the unique format and character of SpringerBriefs (50 to 125 pages), the volumes are compact and easily digestible. Briefs allow authors to present their ideas and readers to absorb them with minimal time investment. Briefs will be published as part of Springer's eBook collection, with millions of users worldwide. In addition, Briefs will be available for individual print and electronic purchase. Briefs are characterized by fast, global electronic dissemination, standard publishing contracts, easy-to-use manuscript preparation and formatting guidelines, and expedited production schedules.

Both solicited and unsolicited manuscripts focusing on food chemistry are considered for publication in this series. Submitted manuscripts will be reviewed and decided by the series editor, Prof. Dr. Salvatore Parisi.

To submit a proposal or request further information, please contact Dr. Sofia Costa, Publishing Editor, via sofia.costa@springer.com or Prof. Dr. Salvatore Parisi, Book Series Editor, via drparisi@inwind.it or drsalparisi5@gmail.com

More information about this subseries at http://www.springer.com/series/11853

Moawiya A. Haddad · Mohammed I. Yamani ·
Saeid M. Abu-Romman · Maher Obeidat

Chemical Profiles of Selected Jordanian Foods

 Springer

Moawiya A. Haddad
Department of Nutrition and Food
Processing
Faculty of Agricultural Technology
Al-Balqa Applied University
Al-Salt, Jordan

Mohammed I. Yamani
Department of Nutrition and Food
Technology
Faculty of Agriculture
University of Jordan
Amman, Jordan

Saeid M. Abu-Romman
Department of Agricultural Biotechnology
Al-Balqa Applied University
Al-Salt, Jordan

Maher Obeidat
Department of Medical Analysis
Faculty of Science
Al-Balqa Applied University
Al-Salt, Jordan

ISSN 2191-5407   ISSN 2191-5415 (electronic)
SpringerBriefs in Molecular Science
ISSN 2199-689X   ISSN 2199-7209 (electronic)
Chemistry of Foods
ISBN 978-3-030-79819-2   ISBN 978-3-030-79820-8 (eBook)
https://doi.org/10.1007/978-3-030-79820-8

© The Author(s), under exclusive license to Springer Nature Switzerland AG 2021
This work is subject to copyright. All rights are solely and exclusively licensed by the Publisher, whether the whole or part of the material is concerned, specifically the rights of translation, reprinting, reuse of illustrations, recitation, broadcasting, reproduction on microfilms or in any other physical way, and transmission or information storage and retrieval, electronic adaptation, computer software, or by similar or dissimilar methodology now known or hereafter developed.
The use of general descriptive names, registered names, trademarks, service marks, etc. in this publication does not imply, even in the absence of a specific statement, that such names are exempt from the relevant protective laws and regulations and therefore free for general use.
The publisher, the authors and the editors are safe to assume that the advice and information in this book are believed to be true and accurate at the date of publication. Neither the publisher nor the authors or the editors give a warranty, expressed or implied, with respect to the material contained herein or for any errors or omissions that may have been made. The publisher remains neutral with regard to jurisdictional claims in published maps and institutional affiliations.

This Springer imprint is published by the registered company Springer Nature Switzerland AG
The registered company address is: Gewerbestrasse 11, 6330 Cham, Switzerland

Contents

Chapter 1
Traditional Foods in Jordan. From Meat Products to Dairy Foods

Abstract The history of traditional foods and beverages in many Asian countries has influenced Mediterranean lifestyle models and the so-called 'Mediterranean Diet' in particular. The localisation of many Middle Eastern dishes and preparations in countries such as Syria, Saudi Arabia, Egypt, Iraq, Lebanon and Jordan, has been often correlated with peculiar features, including also *halal* production methods for meat products. As a consequence, the traditional cuisine in these areas is particularly variegated, from bread and bakery products to meat-based foods, from fish and seafood dishes to vegetable-based products (with the use of different fruits, nuts, rice, but also dairy products). Naturally, such a variety means also implicitly that food and beverage producers are forced to comply with many quality and regulatory requirements, taking also into account the specificity of all traditional foods and beverages. Each product has its own microbiological, chemical and physical profiles, with clear and/or implicit (and probably unknown) advantages and risks by the food-safety viewpoint. The demonstrable evidence of risk assessment in terms of clear and reliable documentation concerning safety, integrity and legal designation of food and beverage products should be available. Also, the evidence of continuous improvement is mandatory. In this ambit, Jordanian foods—as a representation of the more general Middle Eastern Food Tradition—can represent a peculiar case study. Five of these traditional products—or product classes—are briefly introduced in this Chapter by different viewpoints, with peculiar attention to chemical composition, preparation procedures and other features.

Keywords Dairy food · Milk product · Jordan · Middle East · Bakery · Dessert · Mediterranean diet

Abbreviations

GSFA	Codex General Standard of Food Additives
FNCF	Federazione Nazionale dei Chimici e dei Fisici
FAO	Food and Agriculture Organisation of the United Nations
FBO	Food business operator
FSA BIOHAZ Panel	EFSA Panel on Biological Hazards
ITA	Italian Trade Agency

© The Author(s), under exclusive license to Springer Nature Switzerland AG 2021
M. A. Haddad et al., *Chemical Profiles of Selected Jordanian Foods*,
Chemistry of Foods, https://doi.org/10.1007/978-3-030-79820-8_1

MD   Mediterranean Diet
ME   Middle East
SARS-CoV-2   Severe acute respiratory syndrome coronavirus 2

1.1   Middle East and Traditional Food Products

The history of traditional foods and beverages in many Asian countries has influenced Mediterranean lifestyle patterns and the so-called 'Mediterranean Diet' (MD) in particular. The localisation of many Middle Eastern dishes and preparations in countries such as Syria, Saudi Arabia, Egypt, Iraq, Lebanon and Jordan, has been often correlated with peculiar features, including also *halal* production methods for meat products. As a consequence, the traditional cuisine in these areas is particularly variegated, from bread and bakery products to meat-based foods, from fish and seafood dishes to vegetable-based products (with the use of different fruits, nuts, rice, but also dairy products).

Naturally, such a variety means also implicitly that food and beverage producers are forced to comply with many quality and regulatory requirements, taking also into account the specificity of all traditional foods and beverages. Each product has its own microbiological, chemical and physical profiles, with clear and/or implicit (and probably unknown) advantages and risks by the food-safety viewpoint. As an example, the risk of foreign bodies should be carefully evaluated. The demonstrable evidence of risk assessment in terms of clear and reliable documentation concerning safety, integrity and legal designation of food and beverage products should be available. Also, the evidence of continuous improvement (by means of clear standard operative procedures, good manufacturing practices and the execution of corrective/preventive actions against un-avoidable food-related failures) is mandatory.

In this heterogeneous ambit, Jordanian foods can represent a particular case study, as a sub-group of the more general Middle Eastern Food Tradition. Five of these traditional products—or product classes—are examined by different viewpoints in this book, with peculiar attention to chemical composition, preparation procedures and other features. In detail, Chap. 2 is dedicated to a popular fermented dairy drink, *shaneenah*, with relation to chemical characterisation. The third and the fourth Chapters concern *mujaddara*—a product based on the joint use of lentils and rice— and *kebab* products, including the Jordanian version. Finally, Chaps. 5 and 6 are dedicated to the chemical characterisation of Jordaniansweets depend on soft cheeses (a peculiar example is *kunafeh*) and a traditional milk pudding, *muhallabyyah*.

As a general premise, the attention of food and beverage producers and Official Authorities having control tasks on these matters is centred worldwide on the following hygiene and safety topics (Barbieri et al. 2014; Beulens et al. 2005; Delgado et al. 2016a, b, 2017; FAO 2017; Fiorino et al. 2019; Golan et al. 2002–2004a, b; GSFA 2017; Haddad and Parisi 2020a, 2020b; Haddad et al. 2020a, b,

2021; Italian Institute of Packaging 2011; Lauge et al. 2008; Mania et al. 2016a, b–2018; Mitroff et al. 1987; Olsen and Borit 2018; Parisi 2002a–2016–2019; Parisi et al. 2016–2020; Perreten et al. 1997; Phillips 2003; Pisanello 2014; Sheenan 2007a, b; Silva and Malcata 2000; Silvis et al. 2017; Steinka and Parisi 2006; Zanoli and Naspetti 2002; Zhang 2015; Vairo Cavalli et al. 2008), depending on the 'active player' or 'subject:

(a) Public hygiene with concern to the health of consumers (also defined 'food safety), in terms of eradication of elimination of microbiological risks, chemical dangers, detection of foreign substances in foods (metals, plastics materials, wooden materials, nano-chemicals, etc.) and 'other' risks.

(b) Definition, implementation and improvement of regulatory norms, standards and requirements concerning the production, distribution and consumption of food and beverage products, without distinction between pre-packaged items in the ambit of commercial retailers and the so-called 'street foods'.

(c) Definition of reliable durability features concerning foods and beverages, on the basis of most recent scientific opinions (Dongo 2021; FSA BIOHAZ Panel 2020). In particular, this definition clearly circumscribes the temporal ambit of chemical, microbiological and other evaluations all food products are forced to undergo a progressive transformation of their chemical, physical, microbiological and structural features over time; in other words, foods are always subject to change, without exception and irreversibly, with consequent alterations, according to the First Parisi's Law of Food Degradation (FNCF 2020; Parisi 2002a; Pellerito et al. 2019; Srivastava 2019).

(d) Influence of different food processing and preservation techniques on the durability of food products, taking into account that mechanical processes having the purpose of portioning a food or demolishing its structure in some way, can cause a decrease of initial durability without adequate countermeasures (preservation or storage treatments), in accordance with the Second Parisi's Law of Food Degradation (Baglio 2014; Parisi 2002a; Pellerito et al. 2019).The matter of traceability and authenticity (integrity) with reference to globally-supplied food and beverage commodities at present.

(e) Demonstrable studies concerning risk assessment with relation to above-mentioned points and also declaration of food additives, nutrition data, legally-binding names and potentially confusing attributions, etc.

(f) Demonstrable evidence of continuous improvement in all areas of food design, production and distribution.

Because of the possible occurrence of cyclic overproduction periods, the availability of substitute raw materials at the same price of high-quality ingredients and no-related production crises such as the ongoing 'severe acute respiratory syndrome coronavirus 2' (SARS-CoV-2) pandemics, a joint effort by different stakeholders should be needed (Barone and Parisi 2020; Parisi et al. 2020).

In the specific ambit of food science, the interest in polyphenols and other substances with healthy properties (and related claims) has increased in recent years. As a result, many nutritional lifestyles are correlated with 'old' or 'historical' foods because of the claimed equivalence between 'natural' and 'traditional' products. This situation can be currently observed worldwide and in particular when speaking of foods and beverages of vegetable origin (Haddad et al. 2020a, b; Iommi 2021; Issaoui et al. 2020). Anyway, the claimed presence of 'natural' ingredients in modern foods and beverages may be sometimes questionable without adequate and demonstrable evidence from the analytical viewpoint. The importance of required documentation (quality certified systems are always based on the reliable evidence of production processes and all possible documents accompanying the raw materials to their destination as final food products) should be always taken into account. The following factors—safety-related questions, legally-binding designations or commercial advertising messages correlated with 'food fortification', mandatory standards, analytical reports—are extremely important at present (Haddad et al. 2020a, b).

The preference for certain 'poor' products such as 'street foods' has also demonstrated the continuous trend for the research of non-globalised dietary lifestyles and related products. Many of these lifestyles may be also questionable, but the common point is that 'tradition' is often correlated with 'cultural affinity' and 'healthy nutrition at the same time (Pellerito et al. 2019). As a result, different 'street foods' in Europe and also in other Continents can be proposed to consumers with a predictable success, first of all from the economic viewpoint (Barone and Pellerito 2020). The fact that certain foods may contain potentially dangerous substances should be considered at the same time; however, it i a matter of nutritional intakes per day (Barone and Tulumello 2020).

The broad perspective can be also studied in a more specific geographical area with thousand-year-old knowledge such as Middle East (ME). Some research and review have been proposed so far with relation to selected food products in the Jordanian area and it has also been clarified that similar products are part of a broader cultural heritage belonging to many countries in the Middle Eastern area (Haddad et al. 2021). However, many other products could be discussed and this book is dedicated to this aim. It should be also considered that the subdivision of the book concerns also the peculiar typology of food product: in fact, Chaps. 2, 5 and 6 are dedicated to milk-derived products. On the other side, Chap. 3 concerns a recipe based on vegetable ingredients, while the fourth chapter is dedicated to one of the most-known meat-based foods of the Eastern cuisine. It has to be taken into account that our discussion concerns mainly the Jordanian situation, even if many traditions can be found in other ME countries.

1.2   The Jordan Food Processing Sector. An Introduction

At present, an important portion of Jordanian economy relies on food industry businesses (Baabdullah et al. 2019). Historically, the activity of food business operators (FBO) in Jordan has always concerned the supply of food products both at the National and at the International levels. In general, the domestic market is the preferred ground for small FBO. One of the main examples in this ambit is represented by the bakery sector. On the other hand, the recent structural modification of the food chain in Jordan have progressively favoured the birth and the growth of average- and big-sized FBO in Jordan with the aim of supplying both the domestic and the international markets at the same time (Baabdullah et al. 2019; Gregg et al. 2015). However, such a situation has determined the reduction of governmental subsidies (in the ambit of foods) because of the liberalisation of the internal market ('free market') in the 1990s. As a consequence, FBO have experienced a double problem: on the one side, elevated production costs; on the other side, low gains (Ababsa 2014; Baabdullah et al. 2019).

The above-mentioned situation may be hardly understood because food demand in Jordan has increased in the last years. The following factors have determined this trend (Alwan 2006; Baabdullah et al. 2019; Khraishy 2015; Kumar et al. 2011; Loderer et al. 1991; Robbins 1997):

(a) The Jordanian population has progressively increased in size; in addition, dietary lifestyles have evolved, similarly to other ME countries.
(b) Refugees of the Syrian civil war. Jordan hosts around 658,000 registered Syrian refugees however, the real total is estimated at around 1.3 million when those not registered are taken into account.
(c) The high market competition has caused the lowering of food prices. In addition, the enhanced system of food controls for safety and hygiene purposes has to be taken into account.
(d) The Jordanian food demand has experienced high elasticity if compared with proposed prices. In other terms and speaking of microeconomic situations. A defined % variation of proposed prices on the market has often caused a concomitant % variation of food demands where the % variation of demand is > the % variation of prices. In this situation, the demand variation of food products is defined 'elastic' if compared with prices. Should prices get low, the incomes would easily increase. In Jordan, food prices have constantly decreased; as a result, food demand has become strong if compared with the past.

With relation to FBO, the problem is to face food-related challenges both on the domestic and the international ground (Baabdullah et al. 2019). Actually, Jordanian FBO do not have to comply with different standards if compared with other Nations. In general, the market competition places Jordanian FBO in front of European and Asian Players, with the obvious inclusion of ME non-Jordanian FBO. The main markets outside Jordan are essentially the Gulf area, although Jordanian FBO are obtaining good results in other economic areas also. On the one side, the key food products in the above-mentioned markets are (Khraishy 2015):

(a) Export businesses: olive oil, processed meats, fresh vegetable and fruit products, live sheep and other food preparations.
(b) Import businesses: cheeses, pasta, beef, wheat, fish products and non-olive vegetable oils.

It could be interesting to note that certain excellent products in Jordan are in the same category of imported foods, in particular when speaking of meats and cheeses. The Jordanian market experiences a certain degree of competition in these economic areas. The consumeristic preference has a remarkable role in this situation. On the other hand, the following difficulties should be mentioned as important barriers to market penetration, especially when seeing at the International markets (Baabdullah et al. 2019; Gregg et al. 2015):

(a) Quality complaints (including also simple consumers' claims and consequent dissatisfaction, lack of 'brand loyalty', etc.)
(b) Excessive amount of returned commodities
(c) Difficult implementation of voluntary quality standards and mandatory food-safety requirements
(d) Difficult interaction with sales managers and related structures.

As a clear consequence, the aim of Jordanian stakeholders in the food sector is to overthrow immaterial and 'heavy' market barriers for Jordanian FBO highlighting joint efforts at many levels: the food production processes, from initial design to after-purchase consumer services; the improvement of quality and safety/hygiene standards; and the continuous and progressive alignment with voluntary certification systems, where needed and required. On these bases, it could be possible to counterbalance and probably surpass the amount of non-Jordanian foods at least in the meat- and milk-based sectors where a significant number of food and beverage products enter the Country. The current situation of the Jordanian Foods and Beverages sector and correlated export/import flows, is displayed in Fig. 1.1.

In addition, the nutritional profile (chemical composition) of selected foods and beverages could be discussed in this ambit. Figure 1.2 shows a brief summary of four ME products in relation to main nutrients and fibres, on condition that selected nutrient categories are approximately > 13.0%, while fibres are considered 'interesting' if > 7%. The discussion concerning these profiles (with relation to chemical composition) is offered in Chaps. 2–6, taking into account that Fig. 1.2 does not show one of discussed product—*shaneenah*—because of the low amount of classical nutrients (this circumstance depends on servings).

1.3   An Introduction to the Sector of Milk-Based Foods and Beverages in Jordan. Between Tradition and Innovation

The ME tradition of milk-based products should be considered in the broad ambit of the Mediterranean history of foods and beverages. A recent book concerning food traceability of ME and Jordanian food products (Haddad et al. 2021) has discussed main peculiarities of two traditional milk-based foods: *labaneh* (a dairy

The Jordanian F&B Sector

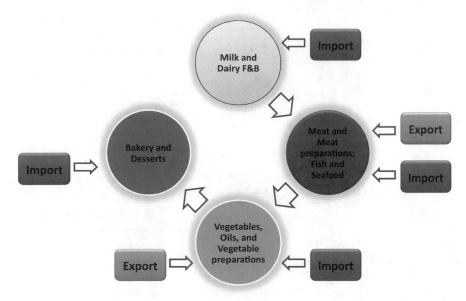

Fig. 1.1 The current situation of the Jordanian Foods and Beverages sector and correlated export/import flows. At present, the main exported food products on the Jordanian market of foods and beverages olive oil, processed meats, fresh vegetable and fruit products, live sheep and other food preparations. On the other side, import businesses concern mainly cheeses, pasta, beef, wheat, fish products and non-olive vegetable oils. From a simplified viewpoint, the four-area subdivided market seems to have always a notable import flow, while export should be improved when speaking of milk and dairy products, typical bakery and dessert foods

product similar to cow's milk yogurt) and *jameed* (a fermented milk product, very popular as ingredient for the Jordanian national dish, *mansaf*) (Hamad et al. 2016–2017). Figure 1.3 shows a general overview of sheep's and goat's milk productions in Jordan such as *shaneenah*, *jameed* and ghee (*samen*).

The first of these foods, *labaneh* (also named *labneh* or *labenah*) the semisolid dairy product made from set yogurt by partial removal of the whey, is widely consumed with olive oil at breakfast and supper meals or as a snack, usually as a sandwich spread (Yamani and Abu-Jaber, 1994; Mihyar et al. 1999); a large proportion of labaneh is still produced by the traditional method of straining set yogurt in cloth bags refers to the Arabic *laban* word (which means 'fermented milk') and its presence may be considered as a cultural heritage of Mediterranean civilizations because similar versions may be found in Greece, Turkey, Spain and the Balkans (Delgado et al. 2017; Abd El-Salam et al. 2011; Rocha et al. 2014; Varnam and Sutherland 1994; Tamime and Robinson 1978; Tamime et al. 1989). Moreover, the importance of *labaneh* is linked to the original raw material: cow's milk and also

Main Nutrients... and Fibres. A brief summary...

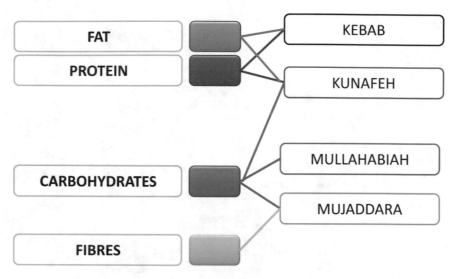

Fig. 1.2 An introduction to the nutritional profile (chemical composition) of selected ME foods and beverages in relation to main nutrients and fibres, on condition that selected nutrient categories are approximately > 13.0%, while fibres are considered 'interesting' if > 7%. The discussion concerning these profiles (with relation to chemical composition) is offered in Chaps. 2–6, taking into account that Fig. 1.2 does not show one of discussed product—*shaneenah*—because of the low amount of classical nutrients

goat's, camel's and sheep's milk can be used (Carod Royo and Sánchez Paniagua 2015; Fuquay et al. 2011).

The second of these milk-based products, *jameed*, is widely found in Jordan and other ME Nations at least (Al-Qudah and Tawalbeh 2011; Buhr 2003; Cheek 2006; Chrysochou et al. 2009; Giraud and Halawany 2006; Golan et al. 2004; Haddad et al. 2021; Jasnen-Vullers et al. 2003; Sodano and Verneau 2004; Starbird and Amanor-Boadu 2006; Verbeke et al. 2007). The tradition links this food to the Bedouins: *jameed* is substantially a preserved and condensed milk and its spherical shape can be considered as a method for easily preserve and store milk- and non-milk-based products, such as Sicilian rice balls of Arabic traditions (Barone and Pellerito 2020). In the *jameed* ambit, sheep's and goat's milk are used, although camel's and cow's milk may be considered as basic raw materials (Haddad et al. 2021).

These products can be used with the aim of representing and discussing briefly the current situation of milk and milk-related industries in Jordan. From the viewpoint of economists, the Jordanian milk sector—and also other ME economical entities—rely on a remarkable quantity of crude raw milk from small farmers, with a consequent low number of average- and big-sized companies in the ambit of

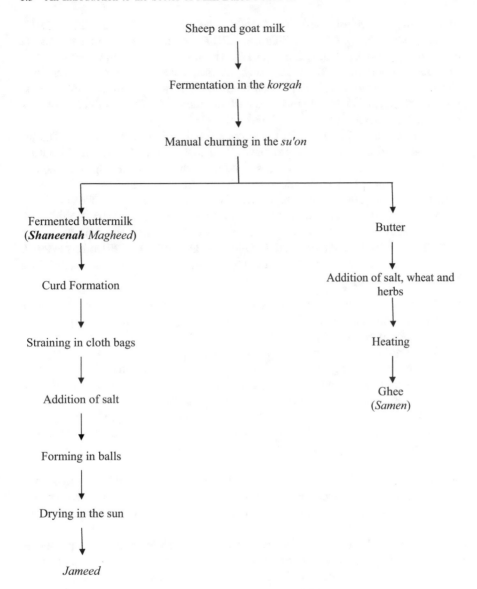

Fig. 1.3 Steps usually followed in the production of the traditional dairy products in Jordan

primary collection (Al Hiary et al. 2013). On the other hand, the number of processors may be lower than initial collectors (milk cooperatives) and this situation is
evident when speaking of *labaneh* processes. The role of household agents and
women has to be considered when speaking of milk collection and microeco-
nomical activities (Al Hiary et al. 2013). The role of women in the sector, in
particular, allows us to consider the main processing activity when speaking of
milk-based products: in general, milk can be used as it is, or processed with the aim

of obtaining buttermilk, pasteurised and boiled cheeses, (Haddad 2015, 2017; Haddad et al 2017; Haddad & Yamani 2017) yogurts and traditional *ghee* (a clarified butter) (Riziq 2017; Zahra 2017). The use of sheep's and goat's milk is extremely important in ME areas, such as the Occupied Palestinian Territory. Consequently, many efforts have been made in these areas with the aim of maximising the yield of milk and processed milk products, with the introduction of milk concentration (obtained food: powdered milk) (Fahmi 2017).

As a result, the milk sector does not concern only *labaneh* and *jameed*. The list of milk-derived and traditional products found in Jordan and other ME countries can at least include the following specialties, where some composite food is mentioned because milk is one of the ingredients (Abu-Jdayil and Mohameed 2004; Abu-Jdayil et al. 2004; Abushihab 2015; Alhammd 2020; Alsharif et al. 2019; Anonymous 2018; Bal and Nath 2006; Bawadi et al. 2009; Elias 1995; Ereifej et al. 2005; Haddad et al. 2020a, b; Khalifa and Shata 2018; Kraig and Sen 2013; ITA 2015; Mazhar 2015; Mutlu et al. 2010; Öğütcü et al. 2017; Oran 2015; Russo 2017; Saad and Ewida 2018; Saleh 2017; Schiff 2017; Shaker 1988; Siçramaz et al. 2016; Soydinç et al. 2016):

(a) *Halva* or *halawa*, a mixed dessert obtained from milk, sesame seed paste (*tahini*), cooked sugar, *Saponaria officinalis* (halva roots) and pistachios (although other minor ingredients such as almonds may be used)

(b) *Muhallabiah, muhallabieh, muhallabiyya,* or *muhallebi* (in Turkey), a milk-based pudding very common in Jordan and in other ME countries (Lebanon, Syria, Egypt and Turkey). Basically and taking into account that each Nation has different versions, this food can be prepared with fresh whole milk, white sugar, rice flour and other minor ingredients (including rose water). One of the possible versions is prepared dissolving corn flour (instead of rice flour) in cold water, then adding it to warm milk and mixing vigorously until boiling and pouring in cups

(c) *Sahlab*, hot-drink milk containing *Orchis anatolica* Boiss tubers (dried powder), cinnamon powder and other optional ingredients. It is a different version of *muhallabyyah* (the original type contains rice or corn starch instead of *O. anatolica* powder)

(d) *Shaneenah, shaneeneh,* or *shenina*, a particular fresh beverage. This product is a goat milk yogurt aged and salted (other ingredients: water and salt) and one of the six main milk-based products in the local economy (milk, yogurt, *labaneh, shaneenah, jameed,* other cheeses and butter). It is also named *doogh* in Persian. Different *shaneenah* versions are available in the ME at least, including the Lebanese type (containing mint)

(e) *Kunafeh, kunafa,knafeh,* or *kanafeh*, a typical Arabic dessert. Basically, it is a sweet product: a flour batter worked with the aim of obtaining vermicelli shapes. The obtained intermediate is sprinkled with thick sugar syrup and subsequently covered with desalted white sheep and goat cheese followed by covering with a thin layer of the vermicelli and heating over low fire until the bottom turns golden brown (Yamani et al, 1997).

The variety of possible milk-based products, dairy foods and derived products which contain milk or cheeses as a basic ingredient do not allow discussing the matter in a single chapter only. This book will discuss chemical features and other characteristics concerning three of the above-mentioned foods: *shaneenah* (Chap. 2), *kunafeh* (Chap. 5) and *muhallabyyah* (Chap. 6).

1.4 The Traditional Meat Sector in Jordan

It has been reported that the Arab dietary lifestyle is correlated with MD (Alsharif et al. 2019; Delgado et al. 2017). In detail, generally available dished in the ME include cheeses (domestic and imported products) (Haddad 2011; Haddad 2015; Haddad et al. 2015; Haddad 2017; Haddad and Yamani 2017, Haddad et al. 2017; Haddad and Abu-Romman 2020), vegetables, composite foods, bread and other cereal-based products, cooked vegetables, processed meats and meat preparations, olive oil and other vegetable fats, different sweets, etc. (Alsharif et al. 2019).

Meat preparations (the exclusion of pork meat has to be always remembered), including composite foods where meat has an important role, have to be considered when speaking of ME diet and specifically in Jordan. The following list shows some of the most important and traditionally linked to popular history foods in the Jordanian ambit at least (Haddad et al. 2021; Tukan et al. 2011):

(1) The National dish, *mansaf* (منسف). Ingredients are: rice; *jameed* soup or yogurt; cooked meat lamb or goat); local bread sheets (*shirak*); butter *ghee* from ewe's milk (local name: *samin baladi*); toppings (fried almonds chopped parsley, pine seeds)

(2) *Shushbarak* (ششبرك) "Athan El-Shaieb"is another local name. Ingredients: Stiff wheat dough; cooked minced meat with onions; *jameed*; butter *ghee* or oil; yogurt; salt; other minor ingredients, including (chickpeas and turmeric powder)

(3) *Kubbeh or Kibbeh*(كبة). Ingredients: burghol (bulgur) and flour; minced meat; onion. Optional ingredients fo*r kubbeh* type: *jameed* or yogurt. Kubbeh is an oval shape bulgur that contains minced cooked meat mixed with chopped cooked onions, sumac spice and pine nuts (tomatoes and pepper may be added). The oval shape finally sealed and fried in vegetable oil to give its popular brown oval shape. It could be eaten alone or with cooked yogurt or as a decoration of popular Jordanian dish ' Mansaf'. With reference to *mansaf*, meat is only one of the basic ingredients. As mentioned recently, the use of *jameed* sauce (from solar-dried buttermilk and salted balls) is peculiar for this recipe. *Shushbarak*or shishbarak is also interesting. (Tukan et al. 2011).

Anyway, being red meat probably more expensive than other meats (example: poultry), the lunch meal should not consider generally meat-based recipes, especially in non-urban areas.

On these bases, it can be affirmed that meat preparations similar to barbecue are preferred in Jordan for public family events and this fact is confirmed when speaking of *kebab*. *Mansaf* is considered in the *Ramadan* fasting and other feasts and often consumed for wedding parties (Tukan et al. 2011).

With reference to the international reputation of *kebab*, it should be also noted that this recipe has spread with interesting numbers when speaking of the Western countries. In detail, there are many *kebab* shops in London (United Kingdom), Berlin (Germany) and other Western countries, meaning that the Western consumer is accustomed and willing to take *kebab* as one of the new ethnic dishes. In general, *kebab* is offered as '*döner kebab*' in form of rotating meat loaves (vertical skewers), while mass retailers can give pre-packaged *kebab* products (Zubaida 2013). However, this type means literally 'rotating *kebab*'; consequently, it is only one of many types, including:

(a) *Döner kebab* (it is served with Arabic bread)
(b) *Dürüm kebab* (it is served with a peculiar Turkish bread, *yufka*)
(c) *Adana kebabı*
(d) *Urfa kebabı*
(e) *Kabab halabi*
(f) *Iskender kebabı*.

All these types are versions and modifications of a traditional kebab food, including some European recipes with pork meat such as the Greek *gyros (γύρος)*. Because of so many types, the discussion concerning *kebab* products in Jordan and other ME versions is shown in Chap. 4.

1.5   The Market of Traditional Vegetable Preparations in Jordan

As above mentioned, the Jordanian Consumer is accustomed to use and prefer vegetable foods and composite products containing vegetables. A recent book has discussed these preferences in detail (Haddad et al. 2020a, b); also, the healthy properties of many fruits and vegetables, including herbal preparations, have to be considered in this ambit with reference to the abundant presence of several active principles such as polyphenols (Laganà et al. 2017–2020).

One of the most characteristic and ancient recipes in the ME is certainly the *mujaddara,majadarah*, or *mejadra* (Kanafani-Zahar 2006; Wilkins and Nadeau 2015). It is a peculiar dish based on lentils and rice (with addition of onions). Actually, the rice/lentils combination can be found in the ME (Lebanon, Jordan, …) and also in Egypt (*kushari*), India (*khichdi*), also because these vegetables have a similar cooling time. Salt, black pepper and olive oil may be added (Rundo 2016; Singh and Singh 2014). Interestingly, *mujaddara* may be also subdivided in two 'brown' or 'yellow' types depending on the lentil typology (green and red lentils, respectively). Because of this situation and also different National versions, the topic concerning *mujaddara* is discussed in detail in Chap. 3.

References

Ababsa M (2014) Jordan food dependency and agriculture sustainability. In: Conference paper, "Developing Agriculture, Cultivating Sovereignty in the Arab Middle-East (1940–2014)?", University of Fribourg, 6–8 November 2014

Abd El-Salam MH, Hippen AR, El-Shafie K, Assem FM, Abbas H, Abd El-Aziz M, Sharaf O, El-Aassar M (2011) Preparation and properties of probiotic concentrated yogurt (labneh) fortified with conjugated linoleic acid. Int J Food Sci Technol 46(10):2103–2110. https://doi.org/10.1111/j.1365-2621.2011.02722.x

Abu-Jdayil B, Mohameed HA (2004) Time-dependent flow properties of starch–milk–sugar pastes. Eur Food Res Technol 218(2):123–127. https://doi.org/10.1007/s00217-003-0829-6

Abu-Jdayil B, Mohameed H, Eassa A (2004) Rheology of starch–milk–sugar systems: effect of heating temperature. Carbohydr Polym 55(3):307–314. https://doi.org/10.1016/j.carbpol.2003.10.006

Abushihab I (2015) Dialect and cultural contact, shift and maintenance among the Jordanians living in Irbid City: a sociolinguistic study. Adv Lang Lit Stud 6(4):84–91

Alhammd Z (2020) Characteristics of dairy value chain in Jordan. As J Econ Bus Account 15 (3):1–9. https://doi.org/10.9734/ajeba/2020/v15i330213

Al Hiary M, Yigezu YA, Rischkowsky B, El-Dine Hilali M, Shdeifat B (2013) Enhancing the dairy processing skills and market access of rural women in Jordan . International Center for Agricultural Research in the Dry Areas (ICARDA) Working Paper No. 565–2016–38922, pp 1–12

Al-Qudah YH, Tawalbeh YH (2011) Influence of production area and type of milk on chemical composition of Jameed in Jordan. J Rad Res Appl Sci 4, 4(B):1263–1270

Alsharif NZ, Khanfar NM, Brennan LF, Chahine EB, Al-Ghananeem AM, Retallick J, Schaalan M, Sarhan N (2019) Cultural sensitivity and global pharmacy engagement in the Arab World. Am J Pharm Educ 83(4):Article 7228

Alwan A (2006) Nutrition in Jordan: update and plan of action. Ministry of Health of Jordan, Amman, and the World Health Organization, Geneva

Anonymous (2018) USAID/LENS Access to Finance (A2F)—purchase order finance—market research in Jordan, June 2018. United States Agency for International Development (USAID), Local Enterprise Support Project, Waghington, D.C

Baabdullah AM, Rana NP, Alalwan AA, Algharabat R, Kizgin H, Al-Weshah GA (2019) Toward a conceptual model for examining the role of social media on social customer relationship management (SCRM) system. In: Elbanna A, Dwivedi Y, Bunker D, Wastell D (eds) Smart Working, Living and Organising. TDIT 2018, Jun 25, Portsmouth, UK. IFIP Advances in Information and Communication Technology 533:102–109. https://doi.org/10.1007/978-3-030-04315-5

Baglio E (ed) (2014) The industry of honey. An introduction. In Chemistry and technology of honey production. SpringerBriefs in Molecular Science. Springer, Cham. https://doi.org/10.1007/978-3-319-65751-6

Bal D, Nath KG (2006) Hazard analysis critical control point in an industrial canteen. Karnataka J Agric Sci 19(1):102–108

Barbieri G, Barone C, Bhagat A, Caruso G, Conley Z, Parisi S (2014) The prediction of shelf life values in function of the chemical composition in soft cheeses. In: The influence of chemistry on new foods and traditional products. SpringerBriefs in Molecular Science. Springer, Cham. https://doi.org/10.1007/978-3-319-11358-6_2

Barone C, Parisi C (2020) The pandemic and curd production. Dairy Ind Int 85(6):28–29
Barone M, Pellerito A (2020) Sicilian street foods and chemistry—the palermo case study. SpringerBriefs in Molecular Science. Springer, Cham. https://doi.org/10.1007/978-3-030-55736-2
Barone M, Tulumello R (2020) Lathyrus sativus: an overview of chemical, biochemical, and nutritional features. In: Lathyrus sativus and Nutrition. SpringerBriefs in Molecular Science. Springer, Cham. https://doi.org/10.1007/978-3-030-59091-8
Bawadi HA, Al-Shwaiyat NM, Tayyem RF, Mekary R, Tuuri G (2009) Developing a food exchange list for Middle Eastern appetisers and desserts commonly consumed in Jordan. Nutr Diet 66(1):20–26. https://doi.org/10.1111/j.1747-0080.2008.01313.x
Beulens AJM, Broens DF, Folstar P, Hofstede GJ (2005) Food safety and transparency in food chains and networks relationships and challenges. Food Control 16(6):481–486. https://doi.org/10.1016/j.foodcont.2003.10.010
Buhr BL (2003) Traceability and information technology in the meat supply chain implications for firm organization and market structure. J Food Distrib Res 34(3):13–26. https://doi.org/10.22004/ag.econ.27057
Carod Royo M, Sánchez Paniagua L (2015) Estudio del efecto de aditivos en la calidad de un snack a base de labneh. Facultad de Veterinaria, Universidad Zaragoza. Available https://zaguan.unizar.es/record/37002/files/TAZ-TFG-2015-3990.pdf. Accessed 30 Sept 2020
Cheek P (2006) Factors impacting the acceptance of traceability in the food supply chain in the United States of America. Rev Sci Tech off Int Epiz 25(1):313–319
Chrysochou P, Chryssochoidis G, Kehagia O (2009) Traceability information carriers. The technology backgrounds and consumers' perceptions of the technological solutions. Appet 53 (3):322–331. https://doi.org/10.1016/j.appet.2009.07.011
Delgado AM, Almeida MDV, Parisi S (2017) Chemistry of the Mediterranean Diet. SpringerBriefs in Molecular Science. Springer, Cham. https://doi.org/10.1007/978-3-319-29370-7
Delgado AM, Vaz de Almeida MD, Parisi S (2016a) Chemistry of the Mediterranean Diet. SpringerBriefs in Molecular Science. Springer, Cham. doi: https://doi.org/10.1007/978-3-319-29370-7
Delgado AM, Vaz de Almeida MD, Barone C, Parisi S (2016b) Leguminosas na dieta mediterrânica—nutrição, Segurança, Sustentabilidade. CISA—VIII Conferência de Inovação e Segurança Alimentar, ESTM—IPLeiria, Peniche, Portugal
Dongo D (2021) Data di scadenza e TMC, linee guida EFSA per la riduzione degli sprechi alimentari. Great Italian Food Trade. Available https://www.greatitalianfoodtrade.it/etichette/data-di-scadenza-e-tmc-linee-guida-efsa-per-la-riduzione-degli-sprechi-alimentari. Accessed 05 Feb 2021
Elias EM (1995) Durum wheat products. In: Di Fonzo N, Kaan F, Nachit M (eds) Durum wheat quality in the Mediterranean region. Options Méditerranéennes: Série A 22:23–31. Centre International de Hautes études agronomiques méditerranéennes (CIHEAM), Zaragoza
Ereifej KI, Rababah TM, Al-Rababah MA (2005) Quality attributes of halva by utilization of proteins, non-hydrogenated palm oil, emulsifiers, gum arabic, sucrose, and calcium chloride. Int J Food Prop 8(3):415–422. https://doi.org/10.1080/10942910500267323
Fahmi E (2017) Alternative milk for newborn sheep. In: Eggens L, Chavez-Tafur J, Pasiecznik N (eds) (2017) Growing hope in Jordan, the Occupied Palestinian Territory and Egypt—Stories from the field, pp. 24–27. Project: Food Security Governance of Bedouin Pastoralist Groups in the Mashreq (NEAR-TS/2012/304–524). Oxfam Italy, Jerusalem
FAO (2017) Food traceability guidance. Food and Agriculture Organization of the United Nations (FAO), Rome.Available http://www.fao.org/3/a-i7665e.pdf. Accessed 29 Sept 2020
Fiorino M, Barone C, Barone M, Mason M, Bhagat A (2019) The intentional adulteration in foods and quality management systems: chemical aspects. Quality systems in the food industry. Springer International Publishing, Cham, pp 29–37
FNCF (2020) La Prima Legge della degradazione alimenti prende il nome da un Chimico italiano. Federazione Nazionale dei Chimici e dei Fisici (FNCF), Rome. Available https://www.chimicifisici.it/la-prima-legge-della-degradazione-alimenti-prende-il-nome-da-un-chimico-italiano/. Accessed 05 Feb 2021

FSA Biohaz Panel (EFSA Panel on Biological Hazards), Koutsoumanis K, Allende A, Alvarez-Ordóñez A, Bolton D, Bover-Cid S, Chemaly M, Davies R, De Cesare A, Herman L, Nauta M, Peixe L, Ru G, Simmons M, Skandamis P, Suffredini E, Jacxsens L, Skjerdal T, Da Silva Felicio MT, Hempen M, Messens W, Lindqvist R (2020) Guidance on date marking and relatedfood information: part 1 (date marking). EFSA J 18(12):6306–6380. https://doi.org/10.2903/j.efsa.2020.6306

Fuquay JW, Fox PF, McSweeney PLH (2011) Milk lipids. Encyclopedia of dairy sciences, vol 3, 2nd edn. Academic Press, Oxford, pp 649–740

Giraud G, Halawany R (2006) Consumers' perception of food traceability in Europe. Proceedings of the 98th EAAE seminar, marketing dynamics within the global Trading System, Chania, Greece

Golan E, Krissoff B, Kuchler F (2002) Traceability for food marketing and food safety: what's the next step? Agric Outlook 288:21–25

Golan E, Krissoff B, Kuchler F (2004a) Food traceability one ingredient in safe efficient food supply. Amber Waves 2(2):14–21

Golan EH, Krissoff B, Kuchler F, Calvin L, Nelson K, Price G (2004b) Traceability in the US food supply: economic theory and industry studies (No. 33939). Economic Research Service, United States Department of Agriculture, Agricultural Economic Report No. 830. United States Department of Agriculture, Washington, DC

Gregg C, Nayef M, Tumurchudur B (2015) Skills for trade and economic diversification (STED). Food processing and beverage sector, international labour organization, regional office for Arab States, Beirut. Available https://www.voced.edu.au/content/ngv%3A83316. Accessed 08 Feb 2021

GSFA (2017) Codex general standard of food additives (GSFA) online database. Codex Alimentarius Commission, Rome. Available http://www.fao.org/gsfaonline/index.html; jsessionid=B20A337D098220B1C4BB75A4B8AB3254. Accessed 29 Sept 2020

Haddad MA (2011) Ph.D. Thesis, University of Jordan. "Microbiological quality of soft white cheese produced traditionally in Jordan and study of its use in the production of probiotic s oft white cheese"

Haddad MA (2015) 29th EFFoST international conference, Athens, Greece. Proceedings 625–630, Elsevier, "Production of probiotic whey drink from released liquid whey of Jordanian soft cheeses".

Haddad MA (2017) Viability of Probiotic bacteria during refrigerated storage of commercial probiotic fermented dairy products marketed In Jordan. J Food Res 6(12):75–81

Haddad MA, Abu-Romman S (2020) PCR-based identification of bovine milk used in goat and sheep local dairy products marketed in Jordan. EurAsian J BioSci 14(11):5267–5272

Haddad MA, Al-Qudah MM, Abu-Romman SM, Maher O, El-Qudah J (2017) Development of traditional Jordanian low sodium dairy products. J Agric Sci 9(1):223–230

Haddad MA, Dmour H, Al-Khazaleh JFM, Obeidat M, Al-Abbadi A, Al-Shadaideh AN, Al-mazra'awi MS, Shatnawi MA, Iommi C (2020) Herbs and medicinal plants in Jordan. J AOAC Int 103(4):925–929. https://doi.org/10.1093/jaocint/qsz026

Haddad MA, El-Qudah J, Abu-Romman S, Obeidat M, Iommi C, Jaradat DSM (2020) Phenolics in mediterranean and middle east important fruits. J AOAC Int 103(4):930–934. https://doi.org/10.1093/jaocint/qsz027

Haddad MA, Parisi S (2020) Evolutive profiles of mozzarella and vegan cheese during shelf-life. Dairy Ind Int 85(3):36–38

Haddad MA, Parisi S (2020) The next big HITS. New Food Mag 23(2):4

Haddad MA, Yamani MI, Abu-alruz K (2015) Development of a Probiotic soft white Jordanian cheese Am-Euras. J Agric Environ Sci 15(7):1382–1391

Haddad MA, Yamani MI (2017) Microbiological quality of soft white cheese produced traditionally in Jordan. J Food Process Technol 8(12):706–712

Haddad MA, Yamani MI, Jaradat DMM, Obeidat M, Abu-Romman SM, Parisi S (2021) Food traceability in Jordan. Current perspectives. SpringerBriefs in Molecular Science. Springer, Cham. https://doi.org/10.1007/978-3-030-66820-4

Hamad MN, Ismail MM, El-Menawy RK (2016) Chemical, rheological, microbial and microstructural characteristics of jameed made from sheep, goat and cow buttermilk or skim milk. Am J Food Sci Nutr Res 3(4):46–55

Hamad MNE, Ismail MM, El-Menawy RK (2017) Impact of innovative formson the chemical composition and rheological properties of jameed. J Nutr Health Food Eng 6(1):00189. https://doi.org/10.15406/jnhfe.2017.06.00189

Iommi C (2021) Chemistry and safety of South American Yerba Mate Teas. SpringerBriefs in Molecular Science. Springer, Cham. https://doi.org/10.1007/978-3-030-69614-6

Issaoui M, Delgado AM, Caruso G, Micali M, Barbera M, Atrous H, Ouslati A, Chammem N (2020) Phenols, flavors, and the mediterranean diet. J AOAC Int 103(4):915–924. https://doi.org/10.1093/jaocint/qsz018

ITA (2015) Giordania—indagine di mercato multisettoriale. In : Proceedings of the International Conference 'Le Regioni della Convergenza e la Cooperazione Euro-Mediterranea', Reggio Calabria, 29th January 2015, Reggio Calabria. Italian Trade Agency (ITA), ICE—Agenzia per la promozione all'estero e l'internazionalizzazione delle imprese italiane, Ufficio Partenariato Industriale e Rapporti con gli Organismi Internazionali, Rome

Italian Institute of Packaging (2011) Aspetti analitici a dimostrazione della conformità del food packaging: linee guida. Prove, calcoli, modellazione e altre argomentazioni. Istituto Italiano Imballaggio, Milan

Jasnen-Vullers MH, van Dorp CA, Beulens AJM (2003) Managing traceability information in manufacture. Int J Inf Manag 23(5):395–413. https://doi.org/10.1016/S0268-4012(03)00066-5

Kanafani-Zahar A (2006) Le Carême et le Ramadan: recréer le corps. Un cas libanais. Revue Des Mondes Musulmans Et De La Méditerranée 113–114:287–300

Khalifa M, Shata RR (2018) Mycobiota and aflatoxins B1 and M1 levels in commercial and homemade dairy desserts in Aswan City Egypt. J Adv Vet Res 8(3):43–48

Khraishy M (2015) Market overview and guide to jordanian market requirements. USDA Foreign Agricultural Service office at U.S. Embassy, Amman, Jordan

Kraig B, Sen CT (eds) (2013) Street food around the world: an encyclopedia of food and culture: an encyclopedia of food and culture. ABC-CLIO, LLC, Santa Barbara

Kumar P, Kumar A, Shinoj P, Raju SS (2011) Estimation of demand elasticity for food commodities in India. Agric Econ Res Rev 24(1):1–14. https://doi.org/10.22004/ag.econ.109408

Laganà P, Avventuroso E, Romano G, Gioffré ME, Patanè P, Parisi S, Moscato U, Delia S (2017) Chemistry and hygiene of food additives. SpringerBriefs in Molecular Science. Springer, Cham. https://doi.org/10.1007/978-3-319-57042-6

Laganà P, Coniglio MA, Fiorino M, Delgado AM, Chammen N, Issaoui M, Gambuzza ME, Iommi C, Soraci L, Haddad MA, Delia S (2020) Phenolic substances in foods and anticarcinogenic properties: a public health perspective. J AOAC Int 103(4):935–939. https://doi.org/10.1093/jaocint/qsz028

Lauge A, Sarriegi J, Torres J (2008) The dynamics of crisis lifecycle for emergency management. Proceedings of the 27th international conference of the system dynamic society, 2009. System Dynamic Society, Albuquerque

Loderer C, Cooney JW, Van Drunen LD (1991) The price elasticity of demand for common stock. J Financ 46(2):621–651. https://doi.org/10.1111/j.1540-6261.1991.tb02677.x

Mania I, Barone C, Caruso G, Delgado A, Micali M, Parisi S (2016a) Traceability in the cheesemaking field. The regulatory ambit and practical solutions. Food Qual Mag 3:18–20. ISSN 2336–4602

Mania I, Delgado AM, Barone C, Parisi S (2018) Traceability in the dairy industry in Europe. Springer International Publishing, Heidelberg, Germany

Mania I, Fiorino M, Barone C, Barone M, Parisi S (2016b) Traceability of packaging materials in the cheesemaking field. The EU Regulatory Ambit. Food Packag Bull 25(4&5):11–16

Mazhar M (2015) The impact of Jordanian health care policy on the maternal and reproductive health care seeking behavior of Syrian Refugee women. Independent Study Project (ISP) Collection. 2057. Available https://digitalcollections.sit.edu/isp_collection/2057. Accessed 08 Feb 2021

Mihyar GF, Yousif AK, Yamani MI (1999) Determination of benzoic and sorbic acids in labaneh by high-performance liquid chromatography. J Food Comp Anal 12(1):53–61. https://doi.org/10.1006/jfca.1998.0804

Mitroff II, Shrivastava P, Udwadia FE (1987) Effective crisis management. Acad Manag Execut 1 (3):283–292

Mutlu AG, Kursun O, Kasimoglu A, Dukel M (2010) Determination of aflatoxin M1 levels and antibiotic residues in the traditional Turkish desserts and ice creams consumed in Burdur City center. J Anim Vet Adv 9(15):2035–2037. https://doi.org/10.3923/javaa.2010.2035.2037

Öğütcü M, Arifoğlu N, Yılmaz E (2017) Restriction of oil migration in tahini halva via organogelation. Eur J Lipid Sci Technol 119(9):1600189. https://doi.org/10.1002/ejlt.201600189

Olsen P, Borit M (2018) The components of a food traceability system. Trend Food Sci Technol 29(2):142–150. https://doi.org/10.1016/j.tifs.2012.10.003

Oran SAS (2015) Selected wild aromatic plants in Jordan. Int J Med Plants 108:686–699

Parisi C, Barone C, Omar SS, Sharma RK (2020) Mozzarella cheese, traceability, consumer choices and chemometrics. Dairy Ind Int 85(8):28–29

Parisi S (2002a) I fondamenti del calcolo della data di scadenza degli alimenti: principi ed applicazioni. Ind Aliment 41(417):905–919

Parisi S, Barone C, Sharma RK (2016) Chemistry and food safety in the EU. The rapid alert system for food and feed (RASFF). Springer Briefs in Molecular Science: Chemistry of Foods, Springer

Pellerito A, Dounz-Weigt R, Micali M (2019) Food sharing: chemical evaluation of durable foods. Springer briefs in molecular science. Springer, Cham. doi: https://doi.org/10.1007/978-3-030-27664-5

Perreten V, Schwarz F, Cresta L, Boeglin M, Dasen G, Teuber M (1997) Nature 389(6653):801–802. https://doi.org/10.1038/39767

Phillips I (2003) Does the use of antibiotics in food animals pose a risk to human health? a critical review of published data. J Antimicrob Chemother 53(1):28–52. https://doi.org/10.1093/jac/dkg483

Pisanello D (2014) Chemistry of Foods: EU legal and regulatory approaches. Springerbriefs in chemistry of foods, springer international publishing, Cham

Riziq AA (2017) Bedouin women at the heart of the matter. In: Eggens L, Chavez-Tafur J, Pasiecznik N (eds) (2017) Growing hope in Jordan, the occupied palestinian territory and Egypt—stories from the field. Project: Food Security Governance of Bedouin Pastoralist Groups in the Mashreq (NEAR-TS/2012/304–524). Oxfam Italy, Jerusalem, pp 20–23

Robbins L (1997) On the elasticity of demand for income in terms of effort. In: Howson S (eds) Economic science and political economy, Palgrave Macmillan, London, pp 79–84. https://doi.org/10.1007/978-1-349-12761-0_6

Rocha DMUP, Martins JDFL, Santos TSS, Moreira AVB (2014) Labneh with probiotic properties produced from kefir: development and sensory evaluation. Food Sci Technol 34(4):694–700. https://doi.org/10.1590/1678-457x.6394

Rundo J (2016) Ricette d'Oriente: la cucina ebraica, cristiana e islamica del Medio Oriente in 90 ricette festive. Edizioni Terra Santa, Milan

Russo R (2017) "La porti un..." cedro del Libano: Muhallabia! Il sorriso vien mangiando. Available http://www.ilsorrisovienmangiando.com/2017/07/la-porti-un-cedro-del-libano-muhallabia.html. Accessed 08 Feb 2021

Saad NM, Ewida RM (2018) Incidence of cronobacter sakazakii in dairy-based desserts. J Adv Vet Res 8(2):16–18

Saleh HMY (2017) Unit operation alteration for developing the characteristics of local white cheese. Dissertation, AlQuds University, Jerusalem – Palestine

Schiff JL (2017) Rare and exotic orchids: their nature and cultural significance. Springer International Publishing, Cham. https://doi.org/10.1007/978-3-319-70034-2

Shaker RR (1988) Technological aspects of the manufacture of halloumi cheese. Dissertation, Massey University, Auckland

Sheenan JJ (2007a) Acidification—19, What problems are caused by antibiotic residues in milk?
    In: McSweeney PLH (ed) Cheese problems solved. Woodhead Publishing Limited,
    Cambridge, and CRC Press LLC, Boca Raton
Sheenan JJ (2007b) Salt in cheese - 46, How does NaCl affect the microbiology of cheese? In:
    McSweeney PLH (ed) Cheese problems solved. Woodhead Publishing Limited, Cambridge,
    and CRC Press LLC, Boca Raton
Siçramaz H, Ayar A, Ayar EN (2016) The evaluation of some dietary fiber rich by-products in ice
    creams made from the traditional pudding–kesme muhallebi. J Food Technol 3(2):105–109.
    https://doi.org/10.18488/journal.58/2016.3.2/58.2.105.109
Silva SV, Malcata FX (2000) Action of cardosin A from Cynara humilis on ovine and caprine
    caseinates. J Dairy Res 67(3):449–454. https://doi.org/10.1017/s0022029900004234
Silvis ICJ, van Ruth SM, van der Fels-Klerx HJ, Luning PA (2017) Assessment of food fraud
    vulnerability in the spices chain: an explorative study. Food Control 81:80–87. https://doi.org/
    10.1016/j.foodcont.2017.05.019
Singh KM, Singh A (2014) Lentil in India: an overview. Munich Personal RePEc Archive, MPRA
    Paper No. 59319, 15 pp. Available https://mpra.ub.uni-muenchen.de/59319/1/MPRA_paper_
    59319.pdf. Accessed 09 Feb 2021
Sodano V, Verneau F (2004) Traceability and food safety: public choice and private incentives,
    quality assurance, risk management and environmental control in agriculture and food supply
    networks. Proceedings of the 82nd Seminar of the European Association of Agricultural
    Economists (EAAE), Bonn, Germany, 14–16 May, Volumes A and B
Soydinç H, Başyiğit B, Hayoğlu I (2016) Effect of fruit addition on the quality characteristics of
    tahini halva. Harran Tarım Ve Gıda Bilimleri Dergisi 20(4):266–275
Srivastava PK (2019) Status report on bee keeping and honey processing. MSME—Development
    institute, ministry of micro, small and medium enterprises, Government of India 107, Industrial
    Estate, Kalpi Road, Kanpur-208012. Available http://msmedikanpur.gov.in/cmdatahien/
    reports/diffIndustries/Status%20Report%20on%20Bee%20keeping%20&%20Honey%
    20Processing%202019-2020.pdf. Accessed 05 Feb 2021
Starbird SA, Amanor-Boadu V (2006) Do Inspection and Traceability Provide Incentives for Food
    Safety? J Agric Res Econ 31(1):14–26
Steinka I, Parisi S (2006) The influence of cottage cheese manufacturing technology and packing
    method on the behaviour of micro-flora. Joint Proc, Deutscher Speditions-und Logistikverband
    e.V., Bonn, and Institut für Logistikrecht & Riskmanagement, Bremerhaven
Tukan SK, Takruri HR, Ahmed MN (2011) Food habits and traditional food consumption in the
    Northern Badia of Jordan. J. Saud Soc Food Nutr 6(1):1–20
Vairo Cavalli S, Silva SV, Cimino C, Malcata FX, Priolo N (2008) Hydrolysis of caprine and
    ovine milk proteins, brought about by aspartic peptidases from Silybum marianum flowers.
    Food Chem 106(3):997–1003. https://doi.org/10.1016/j.foodchem.2007.07.015
Varnam AH, Sutherland JP (1994) Milk and milk products. Technology, Chemistry and
    Microbiology. Chapman and Hall, London
Tamime AY, Robinson RK (1978) Some aspects of the production of concentrated yogurt
    (Labaneh) popular in middle east. Milk Sci Int 33:209–212
Tamime AY, Kalab M, Davies G (1989) Rheology and microstructure of strained yogurt (labneh)
    made from cow's milk by three different methods. Food Str 8(1):15
Verbeke W, Frewer LJ, Scholderer J, De Brabander HF (2007) Why consumers behave as they do
    with respect to food safety and risk information. Anal Chim Acta 586(1–2):2–7. https://doi.org/
    10.1016/j.aca.2006.07.065
Wilkins J, Nadeau R (2015) A companion to food in the ancient world. Wiley Blackwell, Wiley,
    Chichester
Yamani MI, Tukan SK, Abu-Tayeh SJ, English LA (1997) Microbiological quality of Kunafa and
    the development of a Hazard Analysis Critical Control Point (HACCP) plan for its production.
    Dairy Food and Environmental Sanitation 17(10):638–643
Yamani, M. I., & Abu-Jaber, M. M. (1994). Yeast flora of labaneh produced by in-bag straining of
    cow milk set yogurt. Journal of Dairy Science, 77(12), 3558–3564

Zanoli R, Naspetti S (2002) Consumer motivations in the purchase of organic food. A means-end approach. British Food J 104(8):643–653

Zhang D (2015) Best practices in food traceability. In: Proceeding ot the institute of food technologists, 10th July–13th July 2015, Chicago, Available http://www.ift.org/gftc/ ~ /media/ GFTC/Events/Best%20Practices%20in%20Food%20Traceability.pdf. Accessed 29 Sept 2020

Zubaida S (2013) The middle east in London 9, 5:5-6. The London Middle East Institute—SOAS, University of London, London

Zahra WA (2017) Marketing farmers' milk, together. In: Eggens L, Chavez-Tafur J, Pasiecznik N (eds) (2017) Growing hope in Jordan, the occupied palestinian territory and Egypt—Stories from the field, pp. 38–41. Project: Food Security Governance of Bedouin Pastoralist Groups in the Mashreq (NEAR-TS/2012/304–524). Oxfam Italy, Jerusalem

Chapter 2
Shaneenah, a Fermented Dairy Drink. Chemical Features

Abstract This Chapter concerns a popular fermented dairy drink, *shaneenah*, with relation to chemical characterisation. The production of milk and milk-derived products in the ME is certainly relevant and the Jordanian situation can be observed and studied in this ambit because of the variety and types/sub-typologies of different foods and beverages. In Jordan, two-thirds of collected milk are from cows, while the remaining part concerns goat and sheep milk. Moreover, milk is often present in Jordanian and Middle Eastern dishes and recipes, as pasteurised milk, milk powder, drinking and eating yogurt, *labaneh* or *labneh*, *shaneenah*, etc., the traditional *ghee* (a clarified butter), typical cheeses (*halloumi*, *jameed*, etc.), and milk-based desserts and ice creams. In particular, *shaneenah*—a salted and aged yogurt, to be served cold—has a chemical composition depending mainly on the goat milk type. Several variations can be observed because of regional/national differences between the used recipes and raw materials. A recent book concerning food traceability of ME and Jordanian foods (Haddad et al. In Food Traceability in Jordan. Current Perspectives. Springer, Cham 2021; Hamad et al.,.Am J Food Sci Nutr Res 3:46–55, 2016–2017) has considered two of the most typical foods in Jordan and the ME:

(a) *Labaneh,* a fermented milk product, which is commonly used in Jordanian breakfasts and breakfast sandwiches.
(b) The *jameed*, a typical dried cheese of the ancient tradition in the ME area, which is often used for the preparation of the Jordanian national dish, mansaf.

Keywords Carbohydrate · Fat · Jordan · Mediterranean diet · Middle East · Protein · *Shaneenah*

Abbreviations

FBO   Food Business Operator
ITA   Italian Trade Agency
MD   Mediterranean Diet
ME   Middle East

© The Author(s), under exclusive license to Springer Nature Switzerland AG 2021
M. A. Haddad et al., *Chemical Profiles of Selected Jordanian Foods*,
Chemistry of Foods, https://doi.org/10.1007/978-3-030-79820-8_2

2.1   Milk-Based Products in the Middle East. Tradition and Innovation

The cultural heritage of ancient Civilisations in the Middle East (ME) is one of the most interesting features of the current perspectives in terms of 'ethnic' foods, not only in the ambit of Arab consumers. The production of milk and milk-derived products in the ME is certainly relevant and the Jordanian situation can be observed and studied in this ambit because of the variety and types/sub-typologies of different foods and beverages.

It has been recently reported that Jordanian enterprises and food business operators (FBO) in the ambit of the milk sector have been able to produce approximately 485,000 metric tons of fresh milk until at least 2017, while the imported amount of milk should correspond approximately to 25% of the total domestic need. In addition, it should be noted that two-thirds of the amount of needed milk are from cows, while the remaining part concerns goat's and sheep's milk (Hundaileh and Fayad 2019). In general, fresh milk is reported to be used for the production of high-valued foods such as cheeses, with a certain preference for sheep's and goat's milk foods when speaking of export activities. Cow's milk-derived cheeses have also some interest in the regional ambit of ME Countries. Anyway, the basic feature is the separation between drinkable milk (pasteurised product) in the domestic (micro-economical) ambit because of the short shelf-life terms, on the one side and durable foods (cheeses, yogurts) for exporting purposes, on the other side. It appears that low durability and the need of strict preservation and storage requirements—the so-called 'cold chain'—are limiting factors when speaking of milk exports.

By the viewpoint of consumers and with relation to traditions, it has to be mentioned that milk is often present in Jordanian and ME dishes and recipes (Anonymous 2017; Rodinson et al. 2001). With specific relation to Jordanian consumers, the specific manufacturing of dairy products aims at the production of the following food and beverage categories (Figure.....) (Fahmi 2017; Hundaileh and Fayad 2019; Riziq 2017; Zahra 2017):

(a) Pasteurised milk
(b) Milk powder
(c) Drinking and eating yogurt (*labaneh* or *labneh*, *shaneenah or shaneeneh*, etc.)
(d) The traditional *ghee* (a clarified butter)
(e) Traditional cheeses (*halloumi*, Akawi, Nabulsi, Boiled & pasteurised baladi and ... *jameed*, etc.)
(f) Milk-based desserts and ice cream.

A recent book concerning food traceability of ME and Jordanian foods Haddad 2011; Haddad 2015; Haddad et al. 2015; Haddad 2017; Haddad and Yamani 2017, Haddad et al. 2017; Haddad and Abu-Romman 2020, (Haddad et al. 2021; Hamad et al. 2016–2017) has considered two of the most typical foods in Jordan and the ME:

(a) *Labaneh,* a fermented milk product, a strained yogurt salted by 0.9% and used for daily breakfast and breakfast sandwiches.
(b) The *Jameed,* a typical dried cheese of the ancient tradition in the ME area, which is often used for the preparation of the Jordanian national dish, *Mansaf.*

Labaneh (other names: *labenah* or *labneh*, derived from the Arab name for 'milk', *laben*), is substantially a semisolid set yogurt obtained after partial removal of acidic whey (Mihyar et al. 1997). Its presence may be considered as a cultural heritage of Mediterranean Civilizations in ME and non-ME Nations (Abd El-Salam et al. 2011; Delgado et al. 2016a,b–2017; El-Gendi 2015; Rocha et al. 2014; Varnam and Sutherland 1994; Tamime 2006; Tamime and Robinson 1978; Tamime et al. 1989). Interestingly, the relation of *labaneh* with its geographical localisation is evident when speaking of initial raw materials, very common and used in the ME: cow, camel and goat milk (Carod Royo and Sánchez Paniagua 2015; Fuquay et al. 2011).

With concern to solid cheeses, *jameed* balls are a typical food in the ME (Al-Qudah and Tawalbeh 2011; Buhr 2003; Cheek 2006; Chrysochou et al. 2009; Giraud and Halawany 2006; Golan et al. 2002–2004; Haddad et al. 2021; Jasnen-Vullers et al. 2003; Sodano and Verneau 2004; Starbird and Amanor-Boadu 2006; Verbeke et al. 2007). In fact, this milk concentrate (preserved and condensed milk balls) is a heritage of the Bedouins and the spherical shape of *jameed* balls: it can be considered as a method for easily preserve and store milk for extended time periods (Barone and Pellerito 2020). In general, basic raw materials are only sheep's and goat's milk, although camel's and cow's milk are also mentioned (Haddad et al. 2021).

The problem in the milk sector is the fragmentation of the available milk collections. In fact, the Jordanian milk sector relies on a notable number of small cooperatives and farmers: the amount of estimated animals for milk collection can range from 5 to 20 in small entities to more than 500 animals in the largest companies. These numbers are also related to the cow milk sector, but goatherds and shepherds have to be considered because of the relevance of non-cow milk-based foods in Jordan. Anyway, it is reported that medium dairy companies may work with approximately 75 farmers by means of annual contracts (Hundaileh and Fayad 2019; Mercy Corps 2017).

As a clear consequence, the availability of milk for food processing options may be challenging enough because of the fragmentation of offered raw materials at the basis of the milk supply chain. This situation and the effect of durability on export activities (with preference for cheeses, processed products and liquid *jameed* if compared with average and high-water containing foods) should be considered at present (Al Hiary et al. 2013; Fahmi 2017; Haddad et al. 2021; Hundaileh and Fayad 2019; Riziq 2017; Zahra 2017). It has been reported that many efforts have been made in Jordan and ME Countries in these areas with the objective of increasing production yields when speaking of processed milk-derived foods, traceability and modern quality certification systems (Fiorino et al. 2019; Haddad and Parisi 2020; Mania et al. 2016; Meuwissen et al. 2003).

However, the milk sector does not concern mainly *jameed* and other cheeses. The list of traditional ME products where milk is mentioned as a basic or additional ingredient can at least include the following foods (Abu-Jdayil et al. 2004a, 2004b; Abushihab 2015; Anonymous 2018; Bal and Nath 2006; Bawadi et al. 2009; Haddad et al. 2021; Alhammd 2020; Elias 1995; Ereifej et al. 2005; Halkman 2015; Khalifa and Shata 2018; Kraig and Sen 2013; ITA 2015; Mazhar 2015; Mutlu et al. 2010; Öğütcü et al. 2017; Oran 2015; Russo 2017; Saad and Ewida 2018; Saleh 2017; Schiff 2017; Shaker 1988; Siçramaz et al. 2016):

(1) *Shaneenah, shenina,* or *shaneeneh,* a fresh beverage (Fig. 2.1)
(2) *Sahlab,* a hot-drink milk containing *Orchis anatolica* Boiss tubers (dried powder), cinnamon powder and other optional ingredients
(3) *Halva* or *halawa,* a mixed dessert obtained from milk, sesame seed paste (*tahini*), *Saponaria officinalis* (halva roots), cooked sugar and pistachios
(4) *Kunafeh, kunafa, knafeh,* or *kanafeh,* a typical Arabic dessert. Basically, it is a bakery product where goat's cheese (or *halloumi* cheese, in Cyprus) is needed
(5) *Muhallabyyah,* a milk-based pudding very common in Jordan and in the ME area. Basically, it is prepared with fresh whole milk, white sugar, rice flour and other minor ingredients.

Fig. 2.1 Shaneenah, shaneeneh, or shenina, is a particular fresh beverage. This product is a goat or sheep milk yogurt aged and salted (other ingredients: water and salt. It is also named doogh in Persian language. Different shaneenah versions are available in the ME, including the Lebanese type (with mint addition)

These products are essentially foods where milk is not the main ingredient with the exception of *shaneenah*. This beverage (Alhammd 2020; Cappers 2018) is a goat milk yogurt aged and salted. Interestingly, it is one of the main beverages when speaking of milk foods in Jordan, with normal milk and yogurts such as *labaneh*. Actually, there are different*shaneenah* versions in the ME.

What about the chemical composition of this salted product? The following Section concerns basic chemical profiles for this food, also with some relation to production features and other National versions.

2.2 Identification and Chemical Profiles of *Shaneenah*

Apparently, there are not many data concerning the chemical composition of *shaneenah* products, under this name and in our knowledge. Taking into account the peculiar recipe for this product and also different versions (Koçak and Avsar 2009), the following information might be supplied on a 100-g nutritional basis, as a simulated approach:

- Nutritional intake (energy, Kcal): 34
- Protein amount: approximately 2 g
- Fat amount: approximately 1–1.5 g
- Carbohydrate amount: approximately between 0 and 3 g (some products may have 0 g, with a slight decrease of energy intakes, depending on brands).

It should be noted that the recommended serving size of*shaneenah*supplied in a container should be approximately 900 g, with the aim of increasing the energy intake to approximately 306 kcal. In fact, the traditional value of this beverage is linked to tradition and its refreshing power for consumers, while the inherent nutritional profile appears poor enough. Only the bioavailable calcium (and salt) contents should be remarkable because of the natural composition of this product. The only mentioned ingredients are:

(a) Goat's milk-made yogurt
(b) Water
(c) Salt.

The natural composition cannot be dissimilar from a salted and aged yogurt. For this reason and taking into account that goat's milk should be aged and the final *shaneenah* should be served cold, the chemical profile is substantially derived from the original raw material, goat milk, before ageing and turning it into a yogurt. With reference to this milk, the chemical composition is reported as follows (Getaneh et al. 2016; Jenness 1980; Park 2005):

- Protein amount: $3.7 \pm 0.6\%$
- Fat amount: $4.6 \pm 1.4\%$
- Lactose amount: $4.9 \pm 0.6\%$

It should be taken into account that goat's milk may have different compositions depending on various factors, including feeding, geographical locations and involved animals. However, it may be assumed that protein and lactose quantities are constant enough (standard deviations are only 0.6% per 100 ml of goat milk). On the other side, fat amounts can vary notably (for an acceptable confidence level such as K = 2, fat is between 1.8 and 7.4% per 100 ml of goat's milk). On these bases, the composition of *shaneenah* is substantially dependent on the amount of added water and salt, as shown in Table 2.1. In particular, the 'dilution factor' (correspondent to the ratio between the final quantity of *shaneenah* and the original milk amount, D_f) may be indirectly calculated as the mean of ratios between original and final quantities of fats, protein and carbohydrates and assuming that density is roughly similar (even if this assumption is not correct).

As shown in Table 2.1, three dilution factors have been calculated for fat, protein and lactose, respectively. The average D_f from these results is = 2.7 ± 1.4%. Naturally, the obtained result cannot be considered as a good estimation because standard deviation would easily give a minimum and negative value (−0.1%) at the K = 2 confidence level. This situation is surely caused by the extended incertitude concerning fat amounts (the related D_f value is extremely high). The removal of the fat-related dilution factor gives a more acceptable D_f of 1.8 ± 0.2%. It means that 100 ml of original goat milk should be mixed with approximately 40–120 ml of 'salt + water' (average value: 80 ml) when speaking of K = 2 as confidence level. (Table 2.1). Unfortunately, these findings depend mainly on the variability of recipes and associated (theoretical and calculated) compositions. A graphical interpretation of above-mentioned results is displayed in Fig. 2.2.

Some confirmation becomes from the observed and reported data concerning similar products outside Jordan (and extremely popular in the ME). The most similar *shaneenah* products are undoubtedly found in Iran (*doogh*), Turkey and Lebanon (*ayran*), although this yogurt-derived product is also known in India, Macedonia, Armenia, Pakistan, etc. (Halici 2001; Jacobson and Weiner 2008; Tamime 2006). With concern to Persian language, the *doogh* beverage (Arabic name: *ayran*, شنينة, or *dhallë*) is well known, although other names are available. Anyway, the product is similar: it is an aged and salted yogurt and it is served cold (Belal et al. 2019; Otles and Nakilcioglu-Tas 2019; Ministry of Agriculture and Rural Affairs 2001; ZoosNet 2017).

Table 2.1 Calculation of dilution factors for *shaneenah*. The amount of carbohydrates is assumed to be 3.0% (the maximum reported amount) for calculation purposes

Chemical composition, ml/100 ml or g/100 g	Original goat milk	Final *shaheenah*	Dilution factor (D_f)
Fat	4.6	1.0	4.6
Protein	3.7	2.0	1.9
Carbohydrates (mainly lactose)	4.9	3.0	1.6
Global and calculated D_f (on the basis of available data)	–	–	2.7 ± 1.4%

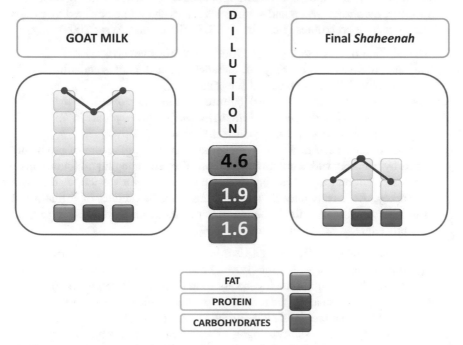

Fig. 2.2 Shaneenah has a main component: goat or sheep milk. The milk composition greatly influences the final chemical profile of the product. Goat's milk may have different compositions depending on various factors. Anyway, protein and lactose quantities are constant enough, while fat amounts can vary notably. On these bases, the composition of *shaneenah* is substantially dependent on the amount of added water and salt, as shown in Table 2.1. The 'dilution factor' (correspondent to the ratio between the final quantity of *shaneenah* and the original milk amount) may be indirectly calculated as the mean of ratios between original and final quantities of fats, protein and carbohydrates and assuming that density is roughly similar (even if this assumption is not correct). Unfortunately, these findings depend mainly on the variability of recipes and associated (theoretical and calculated) compositions.

Several papers and nutritional data are available concerning the *ayran* products. Based on these references and some National requirements, this yogurt has the following features (Koçak and Avsar 2009):

- Nutritional intake (energy, Kcal): 34
- Protein amount: ≥ 2.8 g/ 100 ml
- Fat amount: ≥ 1.8 g/100 ml (full-fat product; other versions from partially skimmed and fat-free milk can be available on the market)
- Lactic acid: ≥ 0.6 g/ 100 ml.

However, there is a variety of homemade and industrial *ayran* products and the same thing is true when speaking of authentic *shaneenah* (or the Lebanese *ayran* version, with mint). As a consequence, the best strategy is to consider only the preparation method for this product without excessive reference to the National area of interest.

Basically, *ayran/shaneenah* and similar beverages should be obtained by means of four alternative procedures (Belal et al. 2019; Koçak and Avsar 2009):

(a) The addition of water to a yogurt (50% water, 50% yogurt), or
(b) Utilising released liquid whey of Labaneh product after standardisation to adjust total solids for the required for shaneenah.
(c) The addition of a lactic (yogurt) culture to milk (*Lactobacillus delbrueckii* subsp. *bulgaricus/ Streptococcus salivarius subsp. Thermophilus by 75%: 25% respectively*) and incubated to produce the required acidity.
(d) The addition of water to a freshly prepared yogurt, with the aim of producing butter by churning with a traditional method. The remaining liquid (buttermilk), after salt addition and mixing, becomes a particular *ayran* version, well known in Turkey. As a consequence, this beverage is a by-product of a traditional Turkish procedure used for obtaining butter as the main food (*yayık ayran*). In this procedure, water is approximately 50%.

The presence of table salt is needed: this is an additive placed in the gelified yogurt with amounts not exceeding 1.0%. Finally, yogurt preparation has to be mixed until the final result, a homogeneous food (Koçak and Avsar 2009).

Based on these information, the relationship between the composition of original goat milk and the final profile of *shaneenah/ayran* is partially clarified. Because of many possible types and sub-typologies (the industrial production is extremely interesting at present), the best strategy should be always connected to the procedure of production. Jordangoat's and sheep's milk are used for production of non commercial amounts of shaneenah. Commercial bottled shaneenah in Jordan is usually produced from cow's milk. In Turkish versions (as a simple example), it may be produced from cow's and sheep's milk also (and the mixture of different milk may be also allowed) (Koçak and Avsar 2009). Apart this clarification, the main differences (also explaining partially the variety of nutritional intakes) in the production of these foods (especially in the industrial field) concern (Angulo et al. 1993; Athanasiadis et al. 2004; Koroleva 1988):

(a) The use of standardised milk with concern to fat matters (this fact is supposed to be the final proof that fat amount is mainly responsible for differences between several *shaneenah* types)
(b) The homogeneisation of intermediate yogurt
(c) The cooling up to 20 °C after yogurt production
(d) The addition of table salts (until a minimum 8% of total solids is obtained in the mixture, belonging 1–2% salt addition), and the final storage after filling (0–4 ° C).In addition, the use of food additives such as pectins or gelatine for stabilising purposes has to be considered when speaking of high-durability products (Kiani et al. 2010; Koçak and Avsar 2009).

References

Abd El-Salam MH, Hippen AR, El-Shafie K, Assem FM, Abbas H, Abd El-Aziz M, Sharaf O, El-Aassar M (2011) Preparation and properties of probiotic concentrated yoghurt (labneh) fortified with conjugated linoleic acid. Int J Food Sci Technol 46(10):2103–2110. https://doi. org/10.1111/j.1365-2621.2011.02722.x

Abu-Jdayil B, Mohameed H, Eassa A (2004) Rheology of starch–milk–sugar systems: effect of heating temperature. Carbohydr Polym 55(3):307–314. https://doi.org/10.1016/j.carbpol.2003. 10.006

Abu-Jdayil B, Mohameed HA (2004) Time-dependent flow properties of starch–milk–sugar pastes. Eur Food Res Technol 218(2):123–127. https://doi.org/10.1007/s00217-003-0829-6

Abushihab I (2015) Dialect and cultural contact, shift and maintenance among the Jordanians living in Irbid City: a sociolinguistic study. Adv Lang Lit Stud 6(4):84–91

Alhammd Z (2020) Characteristics of dairy value chain in Jordan. As J Econ Bus Account 15 (3):1–9. https://doi.org/10.9734/ajeba/2020/v15i330213

Al Hiary M, Yigezu YA, Rischkowsky B, El-Dine Hilali M, Shdeifat B (2013) Enhancing the dairy processing skills and market access of rural women in Jordan . international center for agricultural research in the dry areas (ICARDA) Working Paper No. 565–2016–38922, pp 1–12

Al-Qudah YH, Tawalbeh YH (2011) Influence of production Area and type of milk on chemical composition of Jameed in Jordan. J Rad Res Appl Sci 4, 4(B):1263–1270

Angulo L, Lopez E, Lema C (1993) Microflora present in kefir grains of the Galician region (North-West of Spain). J Dairy Res 60(2):263–267. https://doi.org/10.1017/ S002202990002759X

Anonymous (2017) Middle Eastern Cuisine. Educaterer India, Bihar. Available http:// educatererindia.com/wp-content/uploads/2017/06/Middle-Eastern-Cuisine.pdf. Accessed 08 Feb 2021

Anonymous (2018) USAID/LENS access to finance (A2F)—purchase order finance—market research in Jordan, June 2018. United States Agency for International Development , Local Enterprise Support Project, Waghington, D.C

Athanasiadis I, Paraskevopoulou A, Blekas G, Kiosseoglou V (2004) Development of a novel beverage by fermentation with kefirgranules effect of various treatments. Biotechnol Progr 20 (4):1091–1095. https://doi.org/10.1021/bp0343458

Bal D, Nath KG (2006) Hazard analysis critical control point in an industrial canteen. Karnataka J Agric Sci 19(1):102–108

Barone M, Pellerito A (2020) Palermo's street foods. The authentic pane e panelle. In: Sicilian street foods and chemistry—the palermo case study. SpringerBriefs in Molecular Science. Springer, Cham. https://doi.org/10.1007/978-3-030-55736-2_5

Bawadi HA, Al-Shwaiyat NM, Tayyem RF, Mekary R, Tuuri G (2009) Developing a food exchange list for Middle Eastern appetisers and desserts commonly consumed in Jordan. Nutr Diet 66(1):20–26. https://doi.org/10.1111/j.1747-0080.2008.01313.x

Belal M, Abulola H, Olaimat AN, Odeh RA, Attlee A, Al-Nabulsi AA, Osaili TM, Al-Shami I, Ayyash MM, Obaid RS (2019) Viability of escherichia coli O157:H7 during fermentation and storage of plain and spiced Ayran. J Hygien Eng Design 29:21–25

Buhr BL (2003) Traceability and information technology in the meat supply chain implications for firm organization and market structure. J Food Distrib Res 34(3):13–26. https://doi.org/10. 22004/ag.econ.27057

Cappers RTJ (2018) Digital atlas of traditional food made from cereals and milk, vol 33. Barkhuis/ University of Groningen Library, Groningen, p 639

Carod Royo M, Sánchez Paniagua L (2015) Estudio del efecto de aditivos en la calidad de un snack a base de labneh. Facultad de Veterinaria, Universidad Zaragoza. Available https:// zaguan.unizar.es/record/37002/files/TAZ-TFG-2015-3990.pdf. Accessed 30 Sept 2020

Cheek P (2006) Factors impacting the acceptance of traceability in the food supply chain in the United States of America. Rev Sci Tech off Int Epiz 25(1):313–319

Chrysochou P, Chryssochoidis G, Kehagia O (2009) Traceability information carriers. The technology backgrounds and consumers' perceptions of the technological solutions. Appet 53 (3):322–331. https://doi.org/10.1016/j.appet.2009.07.011

Delgado AM, Almeida MDV, Parisi S (2017) Chemistry of the mediterranean diet. SpringerBriefs in Molecular Science. Springer, Cham. doi: https://doi.org/10.1007/978-3-319-29370-7

Delgado AM, Vaz de Almeida MD, Barone C, Parisi S (2016a) Leguminosas na dieta mediterrânica—nutrição, segurança, sustentabilidade. CISA—VIII Conferência de Inovação e Segurança Alimentar, ESTM—IPLeiria, Peniche, Portugal

Delgado AM, Vaz de Almeida MD, Parisi S (2016b) Chemistry of the mediterranean diet. SpringerBriefs in Molecular Science. Springer, Cham. https://doi.org/10.1007/978-3-319-29370-7

El-Gendi MMN (2015) Comparative study between the microbiological quality of commercial and homemade labenah. Assiut Vet Med J 61, 147:148–153. Available http://www.aun.edu.eg/journal_files/437_J_547.pdf. Accessed 29 Sept 2020

Elias EM (1995) Durum wheat products. In: Di Fonzo N, Kaan F, Nachit M (eds) Durum wheat quality in the Mediterranean region. Options Méditerranéennes : Série A 22:23–31. Centre International de Hautes études agronomiques méditerranéennes, Zaragoza

Ereifej KI, Rababah TM, Al-Rababah MA (2005) Quality attributes of halva by utilization of proteins, non-hydrogenated palm oil, emulsifiers, gum arabic, sucrose, and calcium chloride. Int J Food Prop 8(3):415–422. https://doi.org/10.1080/10942910500267323

Fahmi E (2017) Alternative milk for newborn sheep. In: Eggens L, Chavez-Tafur J, Pasiecznik N (eds) (2017) Growing hope in Jordan, the occupied palestinian territory and Egypt—stories from the field, pp. 24–27. Project: Food Security Governance of Bedouin Pastoralist Groups in the Mashreq (NEAR-TS/2012/304–524). Oxfam Italy, Jerusalem

Fiorino M, Barone C, Barone M, Mason M, Bhagat A (2019) The intentional adulteration in foods and quality management systems: chemical aspects. Quality systems in the food industry. Springer International Publishing, Cham, pp 29–37

Fuquay JW, Fox PF, McSweeney PLH (2011) Milk lipids. Encyclopedia of dairy sciences, vol 3, 2nd edn. Academic Press, Oxford, pp 649–740

Getaneh G, Mebrat A, Wubie A, Kendie H (2016) Review on goat milk composition and its nutritive value. J Nutr Health Sci 3(4):401–410

Giraud G, Halawany R (2006) Consumers' perception of food traceability in Europe. Proceedings of the 98 EAAE seminar, marketing dynamics within the global trading system, Chania, Greece

Golan E, Krissoff B, Kuchler F (2002) Traceability for food marketing and food safety: what's the next step? Agric Outlook 288:21–25

Golan E, Krissoff B, Kuchler F (2004) Food traceability one ingredient in safe efficient food supply. Amber Waves 2(2):14–21

Haddad MA (2011). Ph.D. Thesis, University of Jordan, "Microbiological quality of soft white cheese produced traditionally in jordan and study of its use in the production of probiotic s oft white cheese"

Haddad MA (2015) 29th EFFoST international conference, Athens, Greece. Proceedings 625–630, Elsevier, "Production of probiotic whey drink from released liquid whey of Jordanian soft cheeses"

Haddad MA (2017) Viability of probiotic bacteria during refrigerated storage of commercial probiotic fermented dairy products marketed in Jordan. J Food Res 6(12):75–81

Haddad MA, Abu-Romman S (2020) PCR-based identification of bovine milk used in goat and sheep local dairy products marketed in Jordan. EurAsian J BioSci 14(11):5267–5272

Haddad MA, Al-Qudah MM, Abu-Romman SM, Maher O, El-Qudah J (2017) Development of traditional Jordanian low sodium dairy products. J Agric Sci 9(1):223–230

Haddad MA, Yamani MI, Abu-alruz K (2015) Development of a probiotic soft white Jordanian cheese Am-Euras. J Agric Environ Sci 15(7):1382–1391

Haddad MA, Yamani MI (2017) Microbiological quality of soft white cheese produced traditionally in Jordan. J Food Process Technol 8(12):706–712

Haddad MA, Parisi S (2020) The next big HITS. New Food Magazine 23(2):4

Haddad MA, Yamani MI, Jaradat DMM, Obeidat M, Abu-Romman SM, Parisi S (2021) Food traceability in Jordan. Current perspectives. SpringerBriefs in Molecular Science. Springer, Cham. https://doi.org/10.1007/978-3-030-66820-4

Halici N (2001) Turkish delights. Gastronomica 1(1):92–93. https://doi.org/10.1525/gfc.2001.1.1. 92

Halkman AK (2015) Gıda konusunda yanliş yönlendirmeler. Gazi Üniversitesi Öğretim Üyeleri Derneği, Akademik Bülten, pp 16–20

Hamad MN, Ismail MM, El-Menawy RK (2016) Chemical, rheological, microbial and microstructural characteristics of jameed made from sheep, goat and cow buttermilk or skim milk. Am J Food Sci Nutr Res 3(4):46–55

Hamad MNE, Ismail MM, El-Menawy RK (2017) Impact of innovative formson the chemical composition and rheological properties of jameed. J Nutr Health Food Eng 6(1):00189. https://doi.org/10.15406/jnhfe.2017.06.00189

Hundaileh ML, Fayad MF (2019) Jordan's food processing sector—analysis and strategy for sectoral improvement. Deutsche Gesellschaft für Internationale Zusammenarbeit (GIZ) GmbH, Bonn and Eschborn

ITA (2015) Giordania—indagine di mercato multisettoriale. In : Proceedings of the International Conference 'Le Regioni della Convergenza e la Cooperazione Euro-Mediterranea', Reggio Calabria, 29th January 2015, Reggio Calabria. Italian Trade Agency (ITA), ICE—Agenzia per la promozione all'estero e l'internazionalizzazione delle imprese italiane, Ufficio Partenariato Industriale e Rapporti con gli Organismi Internazionali, Rome

Jacobson S, Weiner D (2008) Yogurt: more than 70 delicious & healthy recipes. Sterling Publishing Company Inc., New York

Jasnen-Vullers MH, van Dorp CA, Beulens AJM (2003) Managing traceability information in manufacture. Int J Inf Manag 23(5):395–413. https://doi.org/10.1016/S0268-4012(03)00066-5

Jenness R (1980) Composition and characteristics of goat milk: review 1968–1979. J Dairy Sci 63 (10):1605–1630. https://doi.org/10.3168/jds.S0022-0302(80)83125-0

Khalifa M, Shata RR (2018) Mycobiota and aflatoxins B1 and M1 levels in commercial and homemade dairy desserts in Aswan City Egypt. J Adv Vet Res 8(3):43–48

Kiani H, Mousavi ME, Razavi H, Morris ER (2010) Effect of gellan, alone and in combination with high-methoxy pectin, on the structure and stability of doogh, a yogurt-based Iranian drink. Food Hydrocoll 24(8):744–754. https://doi.org/10.1016/j.foodhyd.2010.03.016

Koçak C, Avsar YK (2009) Ayran: microbiology and technology. In: Yildiz F (ed) Development and manufacture of yogurt and other functional dairy products, CRC Press, Boca Raton

Koroleva NS (1988) Technology of kefir and kumys. Bull. IDF 227:96–100

Kraig B, Sen CT (eds) (2013) Street food around the world: an encyclopedia of food and culture: an encyclopedia of food and culture. ABC-CLIO, LLC, Santa Barbara

Mania I, Barone C, Caruso G, Delgado A, Micali M, Parisi S (2016) Traceability in the cheesemaking field. The regulatory ambit and practical solutions. Food Qual Mag 3:18–20. ISSN 2336-4602

Mazhar M (2015) The impact of Jordanian health care policy on the maternal and reproductive health care seeking behavior of syrian refugee women. Independent Study Project (ISP) Collection. 2057. Available https://digitalcollections.sit.edu/isp_collection/2057. Accessed 08 Feb 2021

Mercy Corps (2017) Mercy corps: market system assessment for the dairy value chain—irbid & mafraq governorates, Jordan. Mercy Corps, Amman. Available https://data2.unhcr.org/en/documents/download/62006. Accessed 09 Feb 2021

Meuwissen MP, Velthuis AG, Hogeveen H, Huirne R (2003) Traceability and certification in meat supply chains. J Agribus 21(2):167–181

Mihyar GF, Yamani MI, Al-Sa'ed AK (1997) Resistance of yeast flora of labaneh to PS and SB. J Dairy Sci 80(10):2304–2309. https://doi.org/10.3168/jds.S0022-0302(97)76180-0

Ministry of Agriculture and Rural Affairs (2001) Turkish food codex communiqué on fermented milk (Communiqué No: 2001/21). The Official Gazette 3 September 2001, 24512:96–105

Mutlu AG, Kursun O, Kasimoglu A, Dukel M (2010) Determination of aflatoxin m1 levels and antibiotic residues in the traditional Turkish desserts and ice creams consumed in Burdur City Center. J Anim Vet Adv 9(15):2035–2037. https://doi.org/10.3923/javaa.2010.2035.2037

Öğütcü M, Arifoğlu N, Yılmaz E (2017) Restriction of oil migration in tahini halva via organogelation. Eur J Lipid Sci Technol 119(9):1600189. https://doi.org/10.1002/ejlt.201600189

Oran SAS (2015) Selected wild aromatic plants in Jordan. Int J Med Plants 108:686–699

Otles S, Nakilcioglu-Tas E (2019) Nutritional components of some fermented nonalcoholic beverages. In: Grumezescu AM, Holban AM (eds) Fermented beverages, vol 5: The science of beverages. Woodhead Publishing, Cambridge, pp 287–319

Park YW (2005) Goat milk: composition, characteristics. In: Pond WG, Bell AW (eds) Encyclopedia of animal science, 2nd ed., Marcel Dekker, New York

Riziq AA (2017) Bedouin women at the heart of the matter. In: Eggens L, Chavez-Tafur J, Pasiecznik N (eds) (2017) Growing hope in Jordan, the occupied palestinian territory and Egypt—stories from the field. Project: food security governance of bedouin pastoralist groups in the Mashreq (NEAR-TS/2012/304–524). Oxfam Italy, Jerusalem, pp 20–23

Rodinson M, Arberry AJ, Perry C (2001) Medieval arab cookery, Prospect Books, Devon, p 383

Rocha DMUP, Martins JDFL, Santos TSS, Moreira AVB (2014) Labneh with probiotic properties produced from kefir: development and sensory evaluation. Food Sci Technol 34(4):694–700. https://doi.org/10.1590/1678-457x.6394

Russo R (2017) "La porti un..." cedro del Libano: muhallabia! Il sorriso vien mangiando. Available http://www.ilsorrisovienmangiando.com/2017/07/la-porti-un-cedro-del-libano-muhallabia.html. Accessed 08 Feb 2021

Saad NM, Ewida RM (2018) Incidence of Cronobacter sakazakii in Dairy-based Desserts. J Adv Vet Res 8(2):16–18

Saleh HMY (2017) Unit operation alteration for developing the characteristics of local white cheese. dissertation, AlQuds University, Jerusalem—Palestine

Schiff JL (2017) Rare and exotic orchids: their nature and cultural significance. Springer International Publishing, Cham. https://doi.org/10.1007/978-3-319-70034-2

Shaker RR (1988) Technological aspects of the manufacture of halloumi cheese. Dissertation, Massey University, Auckland

Siçramaz H, Ayar A, Ayar EN (2016) The evaluation of some dietary fiber rich by-products in ice creams made from the traditional pudding–kesme muhallebi. J Food Technol 3(2):105–109. https://doi.org/10.18488/journal.58/2016.3.2/58.2.105.109

Sodano V, Verneau F (2004) Traceability and food safety: public choice and private incentives, quality assurance, risk management and environmental control in agriculture and food supply networks. Proceedings of the 82nd Seminar of the European Association of Agricultural Economists, Bonn, Germany, 14–16 May, Volumes A and B

Starbird SA, Amanor-Boadu V (2006) Do inspection and traceability provide incentives for food safety? J Agric Res Econ 31(1):14–26

Tamime AY (2006) Fermented milks. Wiley, Blackwell Science Ltd, Oxford

Tamime AY, Robinson RK (1978) Some aspects of the production of concentrated yogurt (Labaneh) popular in Middle East. Milk Sci Int 33:209–212

Tamime AY, Kalab M, Davies G (1989) Rheology and microstructure of strained yoghurt (labneh) made from cow's milk by three different methods. Food Str 8(1):15

Varnam AH, Sutherland JP (1994) Milk and milk products. Technology, chemistry and microbiology. Chapman and Hall, London

Verbeke W, Frewer LJ, Scholderer J, De Brabander HF (2007) Why consumers behave as they do with respect to food safety and risk information. Anal Chim Acta 586(1–2):2–7. https://doi.org/10.1016/j.aca.2006.07.065

Zahra WA (2017) Marketing farmers' milk, together. In: Eggens L, Chavez-Tafur J, Pasiecznik N (eds) (2017) Growing hope in Jordan, the Occupied palestinian territory and Egypt—stories from the field. Project: Food Security Governance of Bedouin Pastoralist Groups in the Mashreq (NEAR-TS/2012/304–524). Oxfam Italy, Jerusalem, pp 38–41

ZoosNet (2017) How much protein in 1 cup greek yogurt. Protein Choices. Available http://proteinchoices.blogspot.com/2017/06/how-much-protein-in-1-cup-greek-yogurt.html. Accessed 09 Feb 2021

Chapter 3
Rice and Lentils in Jordan. Chemical Profiles of *Mujaddara*

Abstract This Chapter concerns *mujaddara*, a food product based on the joint use of lentils and rice or bulgur in origin, very popular in Jordan and the Middle East. The culinary tradition of vegetable-based foods in the Middle East is correlated with the history of several Asian areas and also the old Europe, especially when speaking of the Middle Age. The so-called 'Mediterranean Diet' has different elements of various origins, often in relation to the concomitant use of selected vegetables and crops. The energy intake of cereals and the notable protein intake of pulses have undoubtedly determined the success of the culinary association. In this ambit, the role of lentils can be studied in detail when speaking of breakfast products, snacks, spreads, soups and also typical Middle Eastern recipes. One of these, in particular, is well-known in the area and in Jordan when speaking of rice/lentils association: the *mujaddara*. The aim of this Chapter is to discuss the chemical profiles of this dish, taking into account that different recipes can give really different results with relation to fat, protein, carbohydrates and fibres.

Keywords Carbohydrate · Fat · Jordan · Mediterranean diet · Middle east · Protein · *Mujaddara*

Abbreviations

FAO   Food and Agriculture Organisation of the United Nations
MD   Mediterranean Diet
ME   Middle East
WHO   World Health Organisation

3.1   Rice-Based Dishes in the Middle East

The culinary tradition of vegetable-based foods in the Middle East (ME) is correlated with the history of several Asian areas and also the old Europe, especially when speaking of the Middle Age. As a result. The so-called 'Mediterranean Diet'

© The Author(s), under exclusive license to Springer Nature Switzerland AG 2021   35
M. A. Haddad et al., *Chemical Profiles of Selected Jordanian Foods*,
Chemistry of Foods, https://doi.org/10.1007/978-3-030-79820-8_3

(MD) has different elements of various origins, often in relation to the use of selected vegetables and crops. It has to be noted that human civilisations have generally followed a dietary pattern based on animal foods, on the one side and vegetable products, on the other side. In the second area of interest, there are several fruits and certain crops which have characterised the history of mankind since the Neolithic Era in Europe and the Indian sub-continent at least (Barone and Tulumello 2020).

This preference is related to the cultivation of cereals on the one side and the use of cereals such as wheat (*Triticum aestivum*) and maize (*Zea mays*) as counter-measures against famines. In fact, cereals are easily storable crops and also high-energy foods (Barone and Tulumello 2020). Consequently, the role of cereal crops as basis or one of the main ingredients for cooked recipes was critical and well-recognised, although their consumption without animal foods was a sign of social distinction between poor and rich classes (Barone and Pellerito 2020).

In addition, the use of non-cereal crops in association with wheat or maize has been reported historically in the old Europe and from the late Roman Empire to the late Middle Age at least. Cereals and grain legumes such as *Lathyrus sativus*, *Lens culinaris*, *Pisum sativum,* and *Vicia faba* have been reported to be cultivated often side-by-side and used together. The energy intake of cereals and the notable protein intake of pulses have undoubtedly determined the success of the culinary associ-ation, in form of soups and other dishes and in different World regions (Abulafia 1978; Alfiero et al. 2017; Bell et al. 2014; de Suremain 2016; Delgado et al. 2017; Heinzelmann 2014; Simopoulus and Bhat 2000; Webb and Hyatt 1988; Wilkins and Hill 2009).

In this ambit, the role of lentils (*L. culinaris*) can be studied in detail when speaking of ME specialities. According to the classification of pulses of the Food and Agriculture Organisation of the United Nations (FAO), lentils become to the *Fabaceae* or *Leguminosae* family (pulses: sub-family: *Faboideae*; tribe: *Lens*). Their use in breakfast products, snacks, spreads, soups and gluten-free foods is often correlated with the remarkable quantity of available protein (20–40%). (Barone and Pellerito 2020; Broomfield 2007; Fernández-Armesto 2002; Pilcher 2017). In addition, ancient legume domestication is reported in the ME at least in Syria, in the *Tell El-Kerkh* site (Barone and Tulumello 2020).

Different ME recipes contain vegetable products as ingredients. One of these, in particular, is well-known in the area and in Jordan when speaking of rice/lentils association: the *mujaddara*. The aim of this Chapter is to discuss the chemical profiles of this ME dish.

3.2 Middle Eastern Dishes. Chemical Profiles of *Mujaddara*

Mujaddara (also named *mejadra* or *mudardara*—we will mainly use this name among available names, although some National difference sometimes may require a diversification) is a peculiar dish (Fig. 3.1) based on lentils, rice or bulgur (burghol) and onions (Al-Khusaibi 2019; FAO 2015; Kanafani-Zahar 2006; Laskar et al. 2019; Salloum 1986). The cereal/non-cereal combination is a well-known feature in the history of foods for human purposes, such as several dishes of the European History (Middle Age). As a recent example, the Sicilian *piciocia* and other Italian recipes are foods based on legumes such as *Lathyrus sativus* and cereals (Kaplan 2008; Montanari 2001). In the same ambit, the ME speciality known as *mujaddara* can be easily found in different Nations (Bisharat 2007). Probably, one of the reasons for the choice of lentils is that these vegetables and rice have similar cooling times (Singh and Singh 2014).

In addition, lentils can supply proteins in a remarkable amount, similar to broad beans and chickpeas (a food easily found in some Traditional Jordanian and Italian products) (Barone and Pellerito 2020; Salloum 1986).Anyway, *mujaddara* is certainly one of the most ancient foods in the ME history: the Persian *Kitāb al-ṭabīkh* (Book of Dishes) cookbook (author: Muhammad bin Hasan al-Baghdadi, year: 1239 A.D.) mentions explicitly the first known recipe for this product (ingredients: rice, lentils and meat) (Paraskiewicz 2020; Sato 2015). Salt, black pepper and olive oil may be added (Rundo 2016).

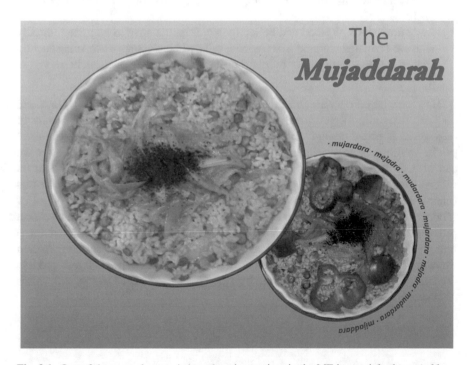

Fig. 3.1 One of the most characteristic and ancient recipes in the ME is certainly the *mujaddara*, *majadarah*, or *mejadra*. It is a peculiar dish based on lentils and rice or bulgur (burghol), with addition of onions. Courtesy of Eng. Abdalla S. Alqawasmi (Bazaria Sweets Factory, Amman, Jordan)

On the basis of several available recipes on the Web, the chemical and nutritional profiles for *mujaddara* (Fig. 3.1) might be calculated (average data) as follows (per one serving, approximately 120–130 g of food) (Amareh 2020; Jawad 2019):

- Energy (Kcal): 281 (minimum, 250; maximum value: 312)
- Protein: 8.5 g (minimum value: 4.0; maximum value: 13.0)
- Fat: 7.5 g (minimum value: 7.0; maximum value: 8.0)
- Saturated fatty acids: 1.0 g
- Carbohydrates: 44.5 g (minimum value: 42.0; maximum value: 47.0)
- Sodium: 444 mg (minimum value: 297; maximum value: 590)
- Fibres: 9.0 g (minimum value: 3.0; maximum value: 15.0).

Actually, different recipes can give really different results, with relation to fat, protein and (naturally) carbohydrate contents, including also the amount of fibres. Basically, the *mujaddara* porridge is obtained simply by mixing lentils with rice or bulgur (burghol) and cooking the whole mass before consumption (Hawtin and Chancellor 1979; World Health Organisation 2012; Zaouali 2012). A simple recipe can be explained as follows (Stefan 2019; Rundo 2016):

(a) Ingredients: lentils, rice or bulgur (burghol), onions, olive oil, salt and black pepper
(b) Basic procedure. Washed and boiled (cooked) lentils have to be mixed initially with pre-browned (caramelised) onions, salt and black pepper and subsequently cooked together with rice or bulgur (burghol). The cooked dish can be served as it is.

However, based on variations, the amount of proteins, carbohydrates, fibres and also salt (as sodium) can greatly vary. On these bases, the nutritional profile of *mujaddara* is broad enough and there are not reliable bases on which a chemical profile could be defined with some valuable result. The only important feature we can note, in this ambit and taking into account that calculated data are based on different recipes for the same name, is that fat matter is approximately constant (please note nutritional data are offered 'per serving' instead of 100 g or ml of product, as usual). This situation explains well the difficult definition of a chemical profile for this recipe, still a matter of culinary art and regional (and international) interpretations worldwide. A graphical interpretation of the chemical composition for *mujaddara* is offered in Fig. 3.2 where data are normalised per 100 g of product (average initial serving is estimated to be 125 g). With reference to Fig. 3.2, salt is obtained as 'sodium (mg) $\times$ 2.5'.

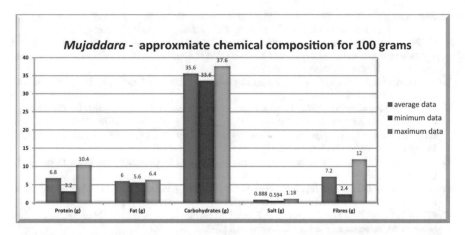

Fig. 3.2 A proposed chemical composition of *mujaddara* per 100 g of product. On the basis of several available recipes on the Web, the chemical and nutritional profiles for *mujaddara* might be calculated (average data) taking into account that available data are generally expressed per serving (between 120 and 130 g of product). Consequently, the nutritional profile of *mujaddara* is broad enough. Fat matter is approximately constant.The present graph shows data after normalisation per 100 g of product (average initial serving is estimated to be 125 g). Salt is obtained as 'sodium (mg) × 2.5'

References

Abulafia D (1978) Pisan commercial colonies and consulates in twelfth-century Sicily. Eng Historical Rev 93(366):68–81

Alfiero S, Giudice AL, Bonadonna A (2017) Street food and innovation: the food truck phenomenon. Brit Food J119(11):2462–2476. https://doi.org/10.1108/BFJ-03-2017-0179

Al-Khusaibi M (2019) Arab traditional foods: preparation, processing and nutrition. In: Al-Khusaibi M, Al-Habsi N, Shafiur Rahman M (eds) Traditional foods, food engineering series, pp. 9–35. Springer International Publishing, Cham. https://doi.org/10.1007/978-3-030-24620-4_2

Amareh H (2020) Mujaddara (Lentils with Brown Rice and Caramelized Onions), March 23, 2020. Amana Nutrition. Available https://amananutrition.com/blog/. Accessed 10 Feb 2021

Barone M, Pellerito A (2020) In: Sicilian street foods and chemistry—the palermo case study. Springerbriefs in molecular science. Springer, Cham. https://doi.org/10.1007/978-3-030-55736-2

Barone M, Tulumello R (2020) Lathyrus sativus: an overview of chemical, biochemical, and nutritional features. In: Lathyrus sativus and nutrition. springerbriefs in molecular science. Springer, Cham. https://doi.org/10.1007/978-3-030-59091-8

Bell JS, Loukaitou-Sideris A (2014) Sidewalk informality: an examination of street vending regulation in China. Int Plan Stud 19(3–4):221–243. https://doi.org/10.1080/13563475.2014.880333

Bisharat G (2007) Talbiyeh days: at villa harun ar-rashid. Jerusalem Q 30:88–98

Broomfield A (2007) Food and cooking in Victorian England: a history. Praeger Publishers, Westport

Delgado AM, Almeida MDV, Parisi S (2017) Chemistry of the mediterranean diet. SpringerBriefs in Molecular Science. Springer, Cham. https://doi.org/10.1007/978-3-319-29370-7

de Suremain CÉ (2016) The never-ending reinvention of 'traditional food'. In: Sébastia B (ed) Eating traditional food: politics, identity and practices. Routledge, Abingdon

FAO (2015) Classification of Commodities (draft). 4. pulses and derived products. food and agriculture organization of the United Nations (FAO), Rome. Available http://www.fao.org/es/faodef/fdef04e.htm. Accessed 10 Feb 2021

Fernández-Armesto F (2002) Near a thousand tables: a history of food. Free Press, Simon and Schuster, New York

Hawtin GC, Chancellor GJ (1979) Food legume improvement and development. In: Proceedings of a workshop held at the University of Aleppo, Syria, 2–7 May 1978, IDRC-126e. The International Center for Agricultural Research in the Dry Areas (ICARDA) and International Development Research Centre (IRCD), Ottawa

Heinzelmann U (2014) Beyond bratwurst: a history of food in Germany. Reaktion Books Ltd., London

Jawad Y (2019) Lebanese Mujadara. Feel Good Foodie. Available https://feelgoodfoodie.net/recipe/mujadara/. Accessed 09th February 2021

Kanafani-Zahar A (2006) Le Carême et le Ramadan: recréer le corps. Un cas libanais. Revue Des Mondes Musulmans Et De La Méditerranée 113–114:287–300

Kaplan L (2008) Legumes in the history of human nutrition. In: Du Bois CM, Tan CB, Mintz SW (eds) The world of soy. University of Illinois Press, Champaign, pp 27–44

Laskar RA, Khan S, Deb CR, Tomlekova N, Wani MR, Raina A, Amin R (2019) Lentil (Lens culinaris Medik.) diversity, cytogenetics and breeding. In: Al-Khayri JM, Jain SM, Johnson DV (eds) Advances in plant breeding strategies: legumes, pp. 319–369. Springer International Publishing, Cham. https://doi.org/10.1007/978-3-030-23400-3_9

Montanari M (2001) Cucina povera, cucina ricca. Quaderni Medievali 52:95–105

Paraskiewicz K (2020) Persian dishes in the 13th century" Kitāb al-ṭabīkh" by al-Baghdādī. In: Michalak-Pikulska B, Piela M, Majtczak T (eds) Oriental languages and civilizations. Jagiellonian University Press, Kraków

Pilcher JM (2017) Planet taco: a global history of Mexican food. Oxford University Press, Oxford

Rundo J (2016) Ricette d'Oriente: la cucina ebraica, cristiana e islamica del Medio Oriente in 90 ricette festive. Edizioni Terra Santa, Milan

Salloum H (1986) Mid-east consumers enlarge market for non-meat foods. Health foods business (USA)

Sato T (2015) 7 Cooking innovations in medieval islam. In: Sugar in the Social Life of Medieval Islam, Brill, Leiden, pp 140–169. https://doi.org/10.1163/9789004281561_009

Simopoulus AP, Bhat RV (2000) Street foods. Karger AG, Basel

Singh KM, Singh A (2014) Lentil in India: an overview. Munich personal RePEc archive, MPRA Paper No. 59319, p 15. Available https://mpra.ub.uni-muenchen.de/59319/1/MPRA_paper_59319.pdf. Accessed 09 Feb 2021

Stefan S (2019) Mujaddara. Bio Salute, Predappio Alta. Available https://www.bio-salute.it/blog/post/37-mujaddara.html. Accessed 10 Feb 2021

Webb RE, Hyatt SA (1988) Haitian street foods and their nutritional contribution to dietary intake. Ecol Food Nutr 21(3):199–209. https://doi.org/10.1080/03670244.1988.9991033

Wilkins J, Hill S (2009) Food in the ancient world. Blackwell Publishing Ltd, Maiden, Oxford, and Carlton

World Health Organization (2012) Promoting a healthy diet for the World Health Organization (WHO) Eastern Mediterranean Region: user-friendly guide. WHO Regional Office for Eastern Mediterranean, Cairo

Zaouali L (2012) Medieval cuisine of the Islamic World : a concise history with 174 recipes. University of California Press, Berkeley

Chapter 4
Chemical and Safety Evaluation of *Kebab*, Including the Jordan Version

Abstract This Chapter concerns a traditional meat preparation in the Middle East, *kebab*. Meat preparations and composite products with an important meat amount have to be discussed when speaking of Middle Eastern foods, such as in Jordan. Three popular dishes can be considered here as some of the most important and traditionally foods linked to popular history foods in the Jordanian ambit: the National Jordanian dish, *mansaf;* the *shushbarak or shishbarak*; and the *kebab*. Actually, this prepared eat is well-known worldwide: in fact, the presence of kebab shops in the main world cities (London, Milan, Berlin, New York, etc.) is synonymous of cultural contamination cosmopolitan integration between different human Civilisations. Another peculiar dish discussed here in relation to Jordan named kubbeh or (*kibbeh*) type, which have many National/Regional versions of the same dish. This Chapter would also briefly discuss current *kebab* products by different perspectives, with reference to chemical profiles.

Keywords Carbohydrate · Fat · *Kebab* · Jordan · Mediterranean diet · Middle east · Protein

Abbreviations

MD Mediterranean Diet
ME Middle East

4.1 Meat Preparations in the Middle East. Reasons for Global Success

The often mentioned relationship between Arab culinary traditions and the Mediterranean Diet (MD) generally include milk-derived foods, composite foods with a qualitative and quantitative presence of vegetables, crops and fruits, olive oil and other vegetable fats, different sweets, etc. (Alsharif et al. 2019; Delgado et al. 2017). The global preference for several Middle East (ME) recipes is now accepted

© The Author(s), under exclusive license to Springer Nature Switzerland AG 2021   41
M. A. Haddad et al., *Chemical Profiles of Selected Jordanian Foods*,
Chemistry of Foods, https://doi.org/10.1007/978-3-030-79820-8_4

and extensively reported, with concern to dishes such as *hummus, taboule,* sun-dried tomatoes *mezze,* etc. (Haddad et al. 2021; Zubaida 2013). In this ambit, the role of processed meats and meat preparations cannot be excluded.

Meat preparations and composite products with an important meat amount have to be discussed when speaking of Middle Eastern foods, such as in Jordan. Naturally, there is the exclusion of pork meat: for this reason, it may be assumed that several ME recipes are imported in non-ME Countries without respect for this important pre-requisite. After this short premise, three popular dishes can be discussed here as some of the most important and traditional foods linked to popular history foods in the Jordanian ambit (Haddad et al. 2021; Haddad 2011, 2015, 2017; Haddad et al. 2015, 2017; Haddad and Yamani 2017; Haddad and Abu-Romman 2020; Tukan et al. 2011):

(1) The National Jordanian dish, *mansaf*
(2) The *shushbarak or shishbarak*
(3) The *kebab.*

Actually, many meat preparations and recipes could be presented in the broader ME ambit. Moreover, the cultural heritage of Arabic Cuisine in non-ME Nations such as Sicily (Italy) should be remembered when speaking of food products containing meat and meat preparations (Barone and Pellerito 2020), meaning the importance of historical spreading of traditions in the old Europe (and the North American Countries, at present) because of migrations from Arabic nations (Bermejo and Sánchez 1998; Capatti and Montanari 2003; Di Fiore 2019; Meri 2005; Wright 1996; Zaouali 2012).

The National Jordanian recipe, *mansaf* (منسف), is prepared with six ingredient categories: rice; *jameed* soup (Al-Qudah and Tawalbeh 2011; Al-Saed et al. 2012; Haddad et al. 2021; Hamad et al. 2016, 2017; Ismail et al. 2017) or yogurt; cooked meat lamb or goat; local bread sheets (*shirak*); butter *ghee* from ewe's milk (local name: *samin baladi*); and toppings (generally, fried almonds, chopped parsley and pine seeds).

The second of mentioned meat dishes, *shushbarak or shishbarak* (ششبرك), is also realised with six ingredients or ingredient categories: stiff wheat dough; cooked minced meat with onions; *jameed*; butter *ghee* or oil; yogurt; salt; and some minor component, including chickpeas and turmeric powder. In this situation and also in the case of *mansaf*, the Jordanian element is evident enough because of the use of *jameed* (generally in form of sauce, because it is a solar-dried preserved fermented milk), a distinctive trait of Jordanian traditions and the Bedouins (Haddad et al. 2021).

Finally, the *kabab* or *kebab* (كباب) should be mentioned. Actually, this prepared meat is well-known worldwide: in fact, the presence of *kebab* shops in the main world cities (London, Milan, Berlin, New York, etc.) is synonymous of cultural contamination and cosmopolitan integration between different human Civilisations (Nassi et al. 2010). Basically, kubbeh or kibbeh dish is discussed here in relation to Jordan as it is used frequently for decoration of Mansaf. Because general ingredients (the stuff)—bulgur (burghol) and flour; minced meat; onion, pine nuts, sumac, spices (tomatoes and pepper may be added)—were cooked before stuffed in the cylindrical oval shape of bulgur mixed with flour. It could be served alone or with cooked yogurt Jameed, demonstrating that the *jameed* speciality can be found in many ME recipes.

With reference to the Jordanian *mansaf*, lamb meat is only one of the basic ingredients and could be found in other ME Countries, while *jameed* sauce (from solar-dried buttermilk formed in salted balls) is peculiar for this recipe in Jordan. The same concept has to be considered when speaking of *shushbarak or shishbarak (Athan El-Shaieb* is another local name). On the other side, *kebab* may be more challenging because of the spreading favour for this food worldwide: as a result, one name and many National/Regional versions of the same dish. This Chapter would briefly discuss current *kebab* products by different perspectives (including the Jordanian viewpoint), with reference to chemical profiles.

4.2 *Kebab* in the Middle East and Everywhere. Chemical Profiles

Meat preparations are especially preferred in Jordan for public family events and this fact is confirmed when speaking of *mansaf* (*Ramadan* fasting dish and other feasts) and *kebab*. Actually, lamb meats are historically more expensive than other meats such as cow and poultry.

Moreover, the *kubbeh or kibbeh* version, in particular, is reported to be often consumed for wedding parties (Tukan et al. 2011). The basic recipe is obtained in the following way: soaked burghol is turned into meat balls with a little amount of flour; obtained balls have to be subsequently filled up with a fried meat (lamb, beef, veal and poultry are used), sumac, spices, pine nuts and onion, in addition, pepper and tomatoes may be added (Kayaardi et al. 2006). Finally, balls have to be sealed and fried On the other side, the *kubbeh or kibbeh* version may be prepared with added *jameed* or yogurt (Tukan et al. 2011).

There are many kebab types worldwide and Jordan is not an exception. The international *kebab* reputation in the Western Countries (United Kingdom, Germany, United States, Italy, etc.) demonstrates clearly that the Western consumer is accustomed and willing to take *kebab* as one of the new ethnic dishes. However, a basic discrimination has to be made when speaking of 'rotating' *kebab*…

In general, *kebab* is offered as Turkish '*döner kebab*' in form of rotating meat loaves (vertical skewers), while mass retailers can give pre-packaged *kebab* products (Zubaida 2013). Consequently, *döner kebab*—generally served with Turkish bread and also named *shawarma, donair,* or *chawarma* (Hosking 2010; Kilic 2009) —concerns substantially 'rotating *kebab*', while several traditional should be mentioned (the list cannot be exhaustive!) (Pandey et al. 2014; Tukan et al. 2011): Although there is a special treatment for producing shawarma in Jordan and the middle east in addition to Turkey, in which the round fresh slices of red meat or from poultry soaked and marinated overnight in special mixed spices and then arranged in layers over each other on a special rod to make a cylindrical shape that broiled with flame just prior serving.

(1) *Dürüm kebab* (it is served with a peculiar Turkish bread, *yufka*)
(2) *Adana kebabı*
(3) *Urfa kebabı*
(4) *Kabab halabi*
(5) *Shami kebab* (popular in India)
(6) *Iskender kebabı*

It has to be clarified that these recipes and *döner kebab* are versions and modifications of a traditional *kebab* food, including some European products with pork meat such as the Greek *gyros* (γύρος).

For these reasons, the definition of a basic recipe and correspondent nutritional analysis in terms of chemical composition at least could be really challenging when speaking of *kebab*. In addition, the analytical detection of (undeclared) pork meat in traditional *kebab* has to be seriously considered as an example of food fraud or adulteration. This possibility should be always considered in the ambit of *halal* foods… modifying necessarily the chemical composition of resulting *kebab* products (Al-Rashedi and Hateem 2016).

On these bases and taking into account the 'international version' (Turkish *döner kebab)* for investigation purposes, it may be inferred on the basis of some studies that (Al-Kutby 2012; Bingöl et al. 2013; Kilic 2009; Liuzzo et al. 2016):

(a) The fat amount varies from 13 to 25 g (on 100 g of product – it could arrive to less than 7 g if stored in air or under vacuum after more than 10 months)
(b) The moisture quantity should be between 43.4 and 69 g (it has to be noted that air-stored sliced*kebab* samples lose water until a possible moisture value of 36.2 g after more than 10 months in frozen state, while under vacuum-packaged slices contain 41.0% of moisture after the same duration)
(c) Protein is approximately between 19 and 26 g
(d) The ash content should be comprised between 1.9 and 3.4%
(e) Salt would be approximately between 1.3 and 2.1%.

A graphical interpretation of above-shown data is presented in Fig. 4.1, taking into account that the graph shows only an average situation. Once more, these values should be considered with care. As a simple example, several researches concerning Italian-prepared *kebab* samples gave the following approximate values (Panozzo et al. 2015): moisture, average value: 59%; protein, 13%; fat, 8.7%; ash, 2.6%; carbohydrates, 12.0%; fibres: 1.2%. As a result, it is really challenging to have reliable data concerning one single *kebab* type… and there are many *kebab* typologies worldwide. Anyway, *kebab* remains one of the main characteristic ME foods sold and consumed in ME and in Jordan.

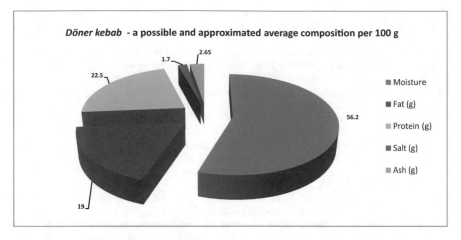

Fig. 4.1 A graphical representation of the 'international version' of Turkish *döner kebab* for investigation purposes. It has to be noted that shown data may vary greatly over time (up to 10 months) if kebab is stored in air or under vacuum, or presented as sliced samples. In detail, average moisture (59.2%) could reach 36.2–41.0% easily, while average fat matter (19.0%) could diminish up to 7%

References

Al-Kutby S (2012) Applications of spice extracts and other hurdles to improve microbial safety and shelf-life of cooked, high fat meat products (doner kebab). Dissertation, University of Plymouth, Plymouth

Al-Qudah YH, Tawalbeh YH (2011) Influence of production area and type of milk on chemical composition of Jameed in Jordan. J Rad Res Appl Sci 4(B):1263–1270

Al-Rashedi NA, Hateem EU (2016) Detection of pork in canned meat using TaqMan Real-time PCR. AL-Muthanna J Pure Sci 3(2):1–6

Al-Saed AK, Al-Groum RM, Al-Dabbas MM (2012) Implementation of hazard analysis critical control point in jameed production. Food Sci Technol Int 18(3):229–239. https://doi.org/10.1177/1082013211427783

Alsharif NZ, Khanfar NM, Brennan LF, Chahine EB, Al-Ghananeem AM, Retallick J, Schaalan M, Sarhan N (2019) Cultural sensitivity and global pharmacy engagement in the arab world. Am J Pharm Educ 83, 4:Article 7228

Barone M, Pellerito A (2020) In: Sicilian street foods and chemistry—the palermo case study. SpringerBriefs in Molecular Science. Springer, Cham. https://doi.org/10.1007/978-3-030-55736-2

Bermejo JEH, Sánchez EG (1998) Economic botany and ethnobotany in Al-Andalus (Iberian Peninsula: Tenth-Fifteenth Centuries), an unknown heritage of mankind. Econ Botan 52(1):15–26

Bingöl EB, Yilmaz F, Muratoğlu K, Bostan K (2013) Effects of vacuum packaging on the quality of frozen cooked döner kebab. Turk J Vet Anim Sci 37(6):712–718

Capatti A, Montanari M (2003) Italian cuisine: a cultural history. Columbia University Press, New York

Delgado AM, Almeida MDV, Parisi S (2017) Chemistry of the mediterranean diet. SpringerBriefs in Molecular Science. Springer, Cham. https://doi.org/10.1007/978-3-319-29370-7

Di Fiore L (2019) Heritage and food history. In: Porciani I (ed) Food heritage and nationalism in Europe. Routledge, London

Haddad MA (2011) Ph.D. Thesis, University of Jordan, "Microbiological quality of soft white cheese produced traditionally in Jordan and study of its use in the production of probiotic s oft white cheese".

Haddad MA (2015) 29th EFFoST international conference, Athens, Greece. Proceedings 625–630, Elsevier, "Production of probiotic whey drink from released liquid whey of Jordanian soft cheeses"

Haddad MA (2017) Viability of probiotic bacteria during refrigerated storage of commercial probiotic fermented dairy products marketed in Jordan. J Food Res 6(12):75–81

Haddad MA, Abu-Romman S (2020) PCR-based identification of bovine milk used in goat and sheep local dairy products marketed in Jordan. EurAsian J BioSci 14(11):5267–5272

Haddad MA, Al-Qudah MM, Abu-Romman SM, Maher O, El-Qudah J (2017) Development of traditional jordanian low sodium dairy products. J Agric Sci 9(1):223–230

Haddad MA, Yamani MI, Abu-alruz K (2015) Development of a probiotic soft white jordanian cheese Am-Euras. J Agric Environ Sci 15(7):1382–1391

Haddad MA, Yamani MI (2017) Microbiological quality of soft white cheese produced traditionally in Jordan. J Food Process Technol 8(12):706–712

Haddad MA, Yamani MI, Jaradat DMM, Obeidat M, Abu-Romman SM, Parisi S (2021) Food traceability in Jordan. Current perspectives. SpringerBriefs in molecular science. Springer, Cham. https://doi.org/10.1007/978-3-030-66820-4

Hamad MN, Ismail MM, El-Menawy RK (2016) Chemical, rheological, microbial and microstructural characteristics of jameed made from sheep, goat and cow buttermilk or skim milk. Am J Food Sci Nutr Res 3(4):46–55

Hamad MNE, Ismail MM, El-Menawy RK (2017) Impact of Innovative formson the chemical composition and rheological properties of jameed. J Nutr Health Food Eng 6(1):00189. https://doi.org/10.15406/jnhfe.2017.06.00189

Hosking R (ed) (2010) Food and language: proceedings of the oxford symposium on food and cooking 2009. Prospect Books, Blackawton

Ismail MM, Farid Hamad MNE, El-Menawy RK (2017) Improvement of chemical properties of jameed by fortification with whey protein. J Nutrition Health Food Sci 6(1):1–11. https://doi.org/10.15226/jnhfs.2018.001117

Kayaardi S, Kundakci A, Kayacier A, Gok V (2006) Sensory and chemical analysis of doner kebab made from turkey meat. J Muscle Foods 17(2):165–173. https://doi.org/10.1111/j.1745-4573.2006.00040.x

Kilic B (2009) Current trends in traditional Turkish meat products and cuisine. LWT-Food Sci Technol 42(10):1581–1589. https://doi.org/10.1016/j.lwt.2009.05.016

Liuzzo G, Rossi R, Giacometti F, Piva S, Serraino A, Mescolini G, Militerno G (2016) Mislabelling of Döner kebab sold in Italy. Italian J Food Saf 5(4):6149. https://doi.org/10.4081/ijfs.2016.6149

Meri JW (ed) (2005) Medieval Islamic civilization: an encyclopedia. Routledge, London

Nassi R, Nuvoloni R, Forzale F, Pedonese F, Gerardo B, Cambi L, D'Ascenzi C (2010) Vendita di Döner kebab nell'area lucchese: risultati dell'attività di sorveglianza. Rivista Italiana Associazione Italiana Veterinari Igienisti 7:55–60

Pandey MC, Harilal PT, Radhakrishna K (2014) Effect of processing conditions on physico-chemical and textural properties of shami kebab. Int Food Res J 21(1):223–228

Panozzo M, Magro L, Erle I, Ferrarini S, Murari R, Novelli E, Masaro S (2015) Nutritional quality of preparations based on Döner Kebab sold in two towns of Veneto Region, Italy: preliminary results. Ital J Food Saf 4(2):4535. https://doi.org/10.4081/ijfs.2015.4535

Tukan SK, Takruri HR, Ahmed MN (2011) Food habits and traditional food consumption in the Northern Badia of Jordan. J. Saud Soc Food Nutr 6(1):1–20

Wright CA (1996) Cucina arabo-sicula and maccharruni. Al-Masāq 9(1):151–177. https://doi.org/
    10.1080/09503119608577029
Zaouali L (2012) Medieval cuisine of the Islamic world: a concise history with 174 recipes.
    University of California Press, Berkeley
Zubaida S (2013) The middle east in London 9, 5:5–6. University of London, London, The
    London Middle East Institute—SOAS

Chapter 5
Jordanian Soft Cheeses. *Kunafeh* and Other Products

Abstract This chapter is dedicated to the chemical characterisation of traditional Jordanian sweets made with cheeses or Nabulsi cheeses, *kunafeh*. The history and origins of Middle Eastern cheeses used in processing this sweet is linked to the Bedouins and their milk production. The differentiation of milk origins could be a problem for local economies when speaking of availability of raw materials because of the notable number of small farmers in the ambit of primary collection and a low number of processors. Available milk types are used with the aim of producing essentially the following food categories: milk for human consumption; buttermilk (*ghee*); cheeses from thermically-treated milk; and yogurt/ yogurt-like types. With relation to cheeses, these products can also be used to obtain foods such as *kunafeh,* a typical Arabic, Jordanian or Nabulsi dessert. This pastry, covered with desalted goat's milk cheese, still appears as the result and the state-of-art of individual cooking masters. Consequently, the chemical profile for this dessert could be questioned due to multiple processing methods and amount and type of fat, while salted cheeses have generally a coherent and reliable composition from the chemical viewpoint.

Keywords Carbohydrate · Fat · *Kunafeh* · Jordan · Mediterranean diet · Middle east · Protein

Abbreviations

ITA Italian Trade Agency
MD   Mediterranean Diet
ME   Middle East

5.1 Milk and Dairy Foods in the Middle East

The tradition of milk-based products gives a distinctive trait to the culinary habits of Mediterranean foods and beverages. In particular, cheeses from cow milk and other milk types can be considered as one of the most ancient methods for preserving milk during extended time periods, with other foods (buttermilk, yogurt, etc.)

© The Author(s), under exclusive license to Springer Nature Switzerland AG 2021
M. A. Haddad et al., *Chemical Profiles of Selected Jordanian Foods*,
Chemistry of Foods, https://doi.org/10.1007/978-3-030-79820-8_5

obtained from milk processing and fermentation (Baglio 2014; Huppertz et al. 2006; Ozen and Dinleyici 2015; Steinkraus 1994). Actually, there are a number of different cheese types and sub-typologies when speaking of cheese: consequently, the ambit of these milk-derived products would be really difficult to be explained and discussed in one book only, even only from the chemical viewpoint.

Anyway, the history of cheese permeates the world of foods and beverages without National boundaries and distinctions. Also, the nature of the original raw material—milk—has to be classified according to the animal origin (cow, cattle, goat, buffalo, camel, etc.). Interestingly, some technical and commercial discrimination can be done on the basis of the correlation between abundance of selected milk types on the one hand and the geographical localisation of cheese production on the other hand. This discrimination has become less important in recent decades because of the commercialisation of different prepared 'curds' as raw materials and related delivery in Countries without a notable abundance and availability of animal milk (Barone and Parisi 2020). Anyway, the relationship between domestic cheese productions and cultural history of a regional area or Country has been often observed, studied and demonstrated so far.

With reference to this book, the Middle East (ME) has a notable number of 'cheese excellences' and the same thing is true when speaking of other milk-derived foods which the spreading has been observed in the Mediterranean area. Chap. 2 has been dedicated to a preliminary examination of the ME milk sector and a recent book concerning food traceability of ME foods (Haddad et al. 2021) has evaluated two of the most know traditional products in Jordan: *labaneh* (a yogurt-like food) and *jameed*, a characteristic fermented milk product, extremely popular as one of the main ingredients for the National Jordanian dish, *mansaf* (Carod Royo and Sánchez Paniagua 2015; Delgado et al. 2016a, b, 2017; El-Gendi 2015; Abd El-Salam et al. 2011; Fuquay et al. 2011; Haddad et al. 2020a, b; Hamad et al. 2016–2017; Mihyar et al. 1999; Rocha et al. 2014; Varnam and Sutherland 1994; Tamime and Robinson 1978; Tamime et al. 1989).

The history and origins of one of these ME foods, *jameed* (balls of preserved and condensed fermented milk under the drying action of sun) is linked to the Bedouins and their sheep and/or goat milk production (Al-Qudah and Tawalbeh 2011; Barone and Pellerito 2020; Haddad et al. 2021; Sodano et al. 2004; Starbird and Amanor-Boadu 2006). Actually, camel's and cow's milk may be used in this ambit (Haddad et al. 2021). Anyway, the differentiation of milk origins could be a problem for local economies when speaking of availability of raw materials. In detail, the ME milk sector—including Jordan—relies on a number of small farmers, while average- and big-sized companies in the ambit of primary collection are a low number (Al Hiary et al. 2013). The problem is that processors are certainly a lower number if compared with farmers, even if the system of milk cooperative has obtained some good result in terms of productivity and economical strength in recent years (Al Hiary et al. 2013; Fahmi 2017). In addition, available milk types are used with the aim of producing essentially the following food categories (Riziq 2017; Zahra 2017):

(a) Milk for human consumption, after adequate sanitising procedures such as pasteurisation, sterilisation and so on. Powdered milk can be included in this group

(b) Buttermilk, including the Fermented buttermilk (Shaneenah)

(c) Clarified *ghee is a* product produced from melting and boiling the sheep and/or goat butter under high temperature to produce a clear oil after evaporating water and addition of bulgur to remove the remaining moisture during boiling and special herbs and spices usually added as an antioxidant source.

(d) Cheeses from thermically-treated milk including pasteurised or boiled soft white cheese.

(e) Yogurt and yogurt-like types (the *labaneh* is one of these foods).

One of the most characteristic ME and Jordanian milk-derived foods has been discussed in Chap. 2 (*shaneenah*, a fermented dairy drink). In addition, a traditional milk pudding—*muhallabyyah*—is discussed in Chap. 6. The following list shows other 'excellences' of the ME traditions, although it has to be noted that some of these products is based on milk with other important non-animal ingredients (Alhammd 2020; Anonymous 2018; Bal and Nath 2006; Halkman 2015; ITA 2015; Mazhar 2015; Mutlu et al. 2010; Saad and Ewida 2018; Siçramaz et al. 2016):

(1) *Halva* or *halawa*, a mixed dessert obtained from milk, sesame seed paste, cooked sugar, *halva* roots and pistachios

(2) *Kunafeh*, a typical Arabic dessert.

In particular, *kunafeh* should be examined in this category in spite of its intrinsic nature: a bakery product. Because of the traditional preparation method and the use of cheese, the chemical composition of *kunafeh* could set aside some alterations.

5.2 *Kunafeh*, Preparation and Chemistry. Nutritional Profiles

Basically, *kunafeh* (also named *kunafa,knafeh, kanafeh,* and *k'nefe bi-jibn*) is a peculiar pastry covered with cheese, well-known in Jordan and Turkey (Yamani et al. 1997; Tamime 2006). It is made by working a flour batter with the aim of obtaining vermicelli shapes. The obtained intermediate is dipped in thick sugar syrup (shredded wheat in syrup) and subsequently covered with desalted goat's milk cheese and/or other ingredients (Seçim and Uçar 2017).

As mentioned before, the presence of a milk-derived ingredient, cheese (similar to *halloumi* cheese, very common in Cyprus), explains the discussion of a composite dessert in this Chapter and with relation to the category of dairy cheeses.

In this ambit, it should be clarified that goat's milk cheese is a boiled white brined cheese. One of the most interesting cheeses in this category and really preferred for *kunafeh*, is *nabulsi* cheese (Humeid 2004): a food where gas holes, openings, rinds and other ruptures should not be observed. Otherwise, the cheese could be defined unfit for the peculiar use, even if it has to be necessarily desalted before using it on the *kunafeh* product (it is a dessert food) (Hejazin and El-Qudah 2009; Humeid et al. 1990). In fact, the role of cheese in the *kunafeh* preparation is the covering of the dessert structure. As a result, the cheese should be highly meltable and stretchable (similarly to Italian *pasta filata* cheeses).

In addition, durability performances should exceed 1 year in freezing under −18 °C and this feature is obliged when speaking of *kunafeh*. So, white brined cheeses, are really needed (Humeid 2004). However, a minimal fermentation can easily compromise these features (Saleh 2017; Tarawneh et al. 2019).

Alternatively, the Jordanian *juben balady* (a soft and unripened white cheese, from ewe and/r goat milk) can be used, on condition that it is refrigerated before use (low durability). In general, it is used for *kunafeh* in spite of low durability performances, provided that microbiological conditions are good enough (low fermentation) (Haddad and Yamani 2017). On the other side, soft white cheeses can be used without preventive salt removal because of the low salt content: brine for this cheese should be only 7%; salt amount in pasteurised cheese should not exceed 10% according to Jordanian standards (Haddad 2011, 2015, 2017; Haddad et al. 2015, 2017; Haddad and Yamani 2017; Haddad and Abu-Romman 2020a). Other traditional boiled cheese in Jordan kept for 6–12 months under about 16% salt. The microbiological problem may be avoided by means of preliminary cooking for cheese. A good way could be the standardisation of procedures and raw materials with the aim of obtaining a moisture value between 46 and 48%, with pH = 6.2–6.8 and protein amount approximately equal to 16 g per 100 g of soft cheese (Haddad et al. 2017). With relation to Turkey, other cheeses such as *antep*, *hatay* cheese, *lor* cheese and *urfa* are used (Seçim and Uçar 2018).

Based on this information, it is clear that nutritional values for used cheeses may be reliable enough when displayed in scientific papers. On the other side, there is a low number of available nutritional information concerning *kunafeh* (although some studies concerning Arabic pastries can be available at present). The problem is that this recipe is still the result and the state-of-art of individual cooking masters. Consequently and based on the obvious diversity of many recipes in the ME at least, it is really challenging to propose a coherent and reliable chemical profile for this dessert. The opposite thing is true concerning cheeses. At present, the following profile for some *kunafeh* recipes (Fig. 5.1) can be supplied for example purposes on the basis of one specific reference (Labban et al. 2020):

- Moisture (grams/100 g): 15.1
- Carbohydrates (grams/100 g): 47.5
- Fat (grams/100 g): 19.7
- Protein (grams/100 g): 16.5
- Ash (grams/100 g): 1.21
- Energy (Kcal/100 g): 433.3.

The *Kunafeh*

Fig. 5.1 *Kunafeh* (also named *kunafa, knafeh, kanafeh* and *k'nefebi-jibn*) is a peculiar pastry covered with cheese, well-known in Jordan and Turkey. It is made by working a flour batter with the aim of obtaining vermicelli shapes. The obtained intermediate is dipped in thick sugar syrup and subsequently covered with desalted goat's milk cheese and other ingredients. Courtesy of Eng. Abdalla S. Alqawasmi (Bazaria Sweets Factory, Amman, Jordan)

Naturally, depending on peculiar product variability and recipes, results can vary. For this reason, above-mentioned data are displayed here as a general average, taking into account a possible variation of $\pm$ 4.0 g for moisture, $\pm$ 2.6 g for fat, $\pm$ 4.1 g for protein and $\pm$ 0.2 for ash.

References

Abd El-Salam MH, Hippen AR, El-Shafie K, Assem FM, Abbas H, Abd El-Aziz M, Sharaf O, El-Aassar M (2011) Preparation and properties of probiotic concentrated yoghurt (labneh) fortified with conjugated linoleic acid. Int J Food Sci Technol 46(10):2103–2110. https://doi.org/10.1111/j.1365-2621.2011.02722.x

Alhammd Z (2020) Characteristics of dairy value chain in Jordan. As J Econ Bus Account 15 (3):1–9. https://doi.org/10.9734/ajeba/2020/v15i330213

Al Hiary M, Yigezu YA, Rischkowsky B, Hilali MED, Shdeifat B (2013) Enhancing the dairy processing skills and market access of rural women in Jordan (565-2016-38922):1–12

Al-Qudah YH, Tawalbeh YH (2011) Influence of production area and type of milk on chemical composition of Jameed in Jordan. J Rad Res Appl Sci 4(B):1263–1270

Anonymous (2018) USAID/LENS Access to finance (A2F)—purchase order finance—market research in Jordan, June 2018. United States agency for international development (USAID), Local Enterprise Support Project, Waghington, D.C

Baglio E (2014) The industry of honey. An introduction. In Baglio E (ed) Chemistry and technology of honey production. SpringerBriefs in Molecular Science. Springer, Cham. https://doi.org/10.1007/978-3-319-65751-6

Bal D, Nath KG (2006) Hazard analysis critical control point in an industrial canteen. Karnataka J Agric Sci 19(1):102–108

Barone C, Parisi C (2020) The pandemic and curd production. Dairy Ind Int 85(6):28–29

Barone M, Pellerito A (2020) Sicilian street foods and chemistry—the palermo case study. SpringerBriefs in Molecular Science. Springer, Cham. https://doi.org/10.1007/978-3-030-55736-2

Carod Royo M, Sánchez Paniagua L (2015) Estudio del efecto de aditivos en la calidad de un snack a base de labneh. Facultad de Veterinaria, Universidad Zaragoza. Available https://zaguan.unizar.es/record/37002/files/TAZ-TFG-2015-3990.pdf. Accessed 30 Sept 2020

Delgado A, Danune JB (2016a) Life-cycle assessment of biodiesel production from microalgae: energy, greenhouse and nutrient balances. Recent advances in microalgal biotechnology, J Published by OMICS Group, CA, USA, pp 89–102

Delgado A, Parisi S, Barone C, Almeida MDV (2016b) Legumes in the mediterranean diet—nutrition, safety, sustainability. Conference on food safety and innovation, Polytechnic Institute of Leiria, 05th May

Delgado AM, Vaz de Almeida MD, Parisi S (2017) Chemistry of the Mediterranean diet. Springer International Publishing, Switzerland

El-Gendi MMN (2015) Comparative study between the microbiological quality of commercial and homemade labenah. Assiut Vet Med J 61(147):148–153. Available http://www.aun.edu.eg/journal_files/437_J_547.pdf. Accessed 29 Sept 2020

Fahmi E (2017) Alternative milk for newborn sheep. In: Eggens L, Chavez-Tafur J, Pasiecznik N (eds) Growing hope in Jordan, the occupied palestinian territory and Egypt—stories from the field. Project: Food Security Governance of Bedouin Pastoralist Groups in the Mashreq (NEAR-TS/2012/304–524). Oxfam Italy, Jerusalem, pp 24–27

Fuquay JW, Fox PF, McSweeney PLH (2011) Milk lipids. Encyclopedia of dairy sciences, vol 3, 2nd edn. Academic Press, Oxford, pp 649–740

Haddad MA (2011) Ph.D Thesis, University of Jordan,. "Microbiological quality of soft white cheese produced traditionally in Jordan and study of its use in the production of probiotic s oft white cheese"

Haddad MA (2015) 29th EFFoST international conference, Athens, Greece. Proceedings 625–630, Elsevier, "Production of probiotic whey drink from released liquid whey of Jordanian soft cheeses"

Haddad MA (2017) Viability of probiotic bacteria during refrigerated storage of commercial probiotic fermented dairy products marketed in Jordan. J Food Res 6(12):75–81

Haddad MA, Abu-Romman S (2020a) PCR-based identification of bovine milk used in goat and sheep local dairy products marketed in Jordan. EurAsian J BioSci 14(11):5267–5272

Haddad MA, El-Qudah J, Abu-Romman S, Obeidat M, Iommi C, Jaradat DMM (2020b) Phenolics in Mediterranean and Middle East important fruits. J AOAC Int 103(4):930–934. Oxford University Press

Haddad MA, Al-Qudah MM, Abu-Romman SM, Maher O, El-Qudah J (2017) Development of traditional jordanian low sodium dairy products. J Agric Sci 9(1):223–230

Haddad MA, Yamani MI, Abu-alruz K (2015) Development of a probiotic soft white jordanian cheese Am-Euras. J Agric Environ Sci 15(7):1382–1391

Haddad MA, Yamani MI (2017) Microbiological quality of soft white cheese produced traditionally in Jordan. J Food Process Technol 8(12):706–712

Haddad MA, Yamani MI, Jaradat DMM, Obeidat M, Abu-Romman SM, Parisi S (2021) Food traceability in Jordan. Current perspectives. SpringerBriefs in Molecular Science. Springer, Cham. https://doi.org/10.1007/978-3-030-66820-4

Halkman AK (2015) Gıda Konusunda Yanlış Yönlendirmeler. Gazi Üniversitesi Öğretim Üyeleri Derneği, Akademik Bülten, pp. 16–20

Hamad MN, Ismail MM, El-Menawy RK (2016) Chemical, rheological, microbial and microstructural characteristics of jameed made from sheep, goat and cow buttermilk or skim milk. Am J Food Sci Nutr Res 3(4):46–55

Hamad MNE, Ismail MM, El-Menawy RK (2017) Impact of innovative formson the chemical composition and rheological properties of Jameed. J Nutr Health Food Eng 6(1):00189. https://doi.org/10.15406/jnhfe.2017.06.00189

Hejazin RK, El-Qudah M (2009) Effect of proteases on meltability and stretchability of Nabulsi cheese. Am J Agric Biol Sci 4(3):173–178

Humeid MA (2004) Investigation on imparting strechability and meltability to white brined Nabulsi cheese. Dissertation, University of Jordan, Amman

Humeid MA, Tukan SK, Yamani MI (1990) In-bag steaming of white brined cheese as a method for preservation. Milchwiss 45(8):513–516

Huppertz T, Upadhyay VK, Kelly AL, Tamime AY (2006) Constituents and properties of milk from different species. In: Tamime AY (ed) Brined cheeses, pp. 1–42. Blackwell Publishing Ltd, Oxford. https://doi.org/10.1002/9780470995860.ch1

ITA (2015) Giordania—indagine di mercato multisettoriale. In: Proceedings of the international conference 'le regioni della convergenza e la cooperazione euro-mediterranea', Reggio Calabria, 29th January 2015, Reggio Calabria. Italian Trade Agency (ITA), ICE—Agenzia per la promozione all'estero e l'internazionalizzazione delle imprese italiane, Ufficio Partenariato Industriale e Rapporti con gli Organismi Internazionali, Rome

Labban L, Thallaj N, Al Masri M (2020) Nutritional value of traditional syrian sweets and their calorie density. J Adv Res Food Sci Nutr 3(1):34–41. https://doi.org/10.24321/2582.3892.202005

Mazhar M (2015) The impact of Jordanian Health care policy on the maternal and reproductive health care seeking behavior of Syrian refugee women. Independent Study Project Collection. 2057. Available https://digitalcollections.sit.edu/isp_collection/2057. Accessed 08 Feb 2021

Mihyar GF, Yousif AK, Yamani MI (1999) Determination of benzoic and sorbic acids in labaneh by high-performance liquid chromatography. J Food Comp Anal 12(1):53–61. https://doi.org/10.1006/jfca.1998.0804

Mutlu AG, Kursun O, Kasimoglu A, Dukel M (2010) Determination of aflatoxin M1 levels and antibiotic residues in the traditional turkish desserts and ice creams consumed in Burdur City Center. J Anim Vet Adv 9(15):2035–2037. https://doi.org/10.3923/javaa.2010.2035.2037

Ozen M, Dinleyici EC (2015) The history of probiotics: the untold story. Benef Microb 6(2):159–165. https://doi.org/10.3920/BM2014.0103

Riziq AA (2017) Bedouin women at the heart of the matter. In: Eggens L, Chavez-Tafur J, Pasiecznik N (eds) (2017) Growing hope in Jordan, the occupied palestinian territory and Egypt—stories from the field, pp. 20–23. Project: Food Security Governance of Bedouin Pastoralist Groups in the Mashreq (NEAR-TS/2012/304–524). Oxfam Italy, Jerusalem

Rocha DMUP, Martins JDFL, Santos TSS, Moreira AVB (2014) Labneh with probiotic properties produced from kefir: development and sensory evaluation. Food Sci Technol 34(4):694–700. https://doi.org/10.1590/1678-457x.6394

Saad NM, Ewida RM (2018) Incidence of cronobacter sakazakii in dairy-based desserts. J Adv Vet Res 8(2):16–18

Saleh HMY (2017) Unit operation alteration for developing the characteristics of local white cheese. Dissertation, AlQuds University, Jerusalem – Palestine

Seçim Y, Uçar G (2017) Evaluation of the desserts; which are hosmerim, cheese halva, kunafah produced in Turkish cuisine-in aspect of tourism. Int J Soc Sci Educ Res 3(5S):1478–1484. https://doi.org/10.24289/ijsser.317678

Seçim Y, Uçar G (2018) Determination of microbiological, chemical, and sensory characteristics of hosmerim desserts derived from sheep, goat, and cow cheese. Euras J Vet Sci 34(3):156–163. https://doi.org/10.15312/EurasianJVetSci.2018.195

Siçramaz H, Ayar A, Ayar EN (2016) The evaluation of some dietary fiber rich by-products in ice creams made from the traditional pudding–kesme muhallebi. J Food Technol 3(2):105–109. https://doi.org/10.18488/journal.58/2016.3.2/58.2.105.109

Sodano V, Verneau F, Schiefer G, Rickert U (2004) Traceability and food safety: public choice and private incentives.

Starbird SA, Amanor-Boadu V (2006) Do inspection and traceability provide incentives for food safety? J Agri and Res Eco 31(1):14–26. Available at SSRN: https://ssrn.com/abstract=902446

Steinkraus KH (1994) Nutritional significance of fermented foods. Food Res Int 27(3):259–267

Tamime AY, Robinson RK (1978) Some aspects of the production of concentrated yogurt (Labaneh) popular in Middle East. Milk Sci Int 33:209–212

Tamime AY, Kalab M, Davies G (1989) Rheology and microstructure of strained Yoghurt (Labneh) made from cow's milk by three different methods. Food Str 8(1):15

Tamime AY (2006) Brined cheeses. Blackwell Publishing Ltd., Oxford

Tarawneh H, Al Ismail K, Haddadin M, Sadder M (2019) Improving the meltability and stretchability of white brined cheese using enzymatic and chemical modifications to produce high-quality kunafa and other popular local sweets and pastries.Int J Appl Nat Sci 8(6):61–66

Yamani MI, Tukan SK, Abu-Tayeh SJ, English LA (1997) Microbiological quality of Kunafa and the development of a hazard analysis critical control point (HACCP) plan for its production. Dairy Food Environ Sanitat 17(10):638–643

Varnam AH, Sutherland JP (1994) Milk and milk products. Technology, chemistry and microbiology. Chapman and Hall, London

Zahra WA (2017) Marketing farmers' milk, together. In: Eggens L, Chavez-Tafur J, Pasiecznik N (eds) Growing hope in Jordan, the occupied palestinian territory and Egypt—stories from the field. Project: Food Security Governance of Bedouin Pastoralist Groups in the Mashreq (NEAR-TS/2012/304–524). Oxfam Italy, Jerusalem, pp 38–41

Chapter 6
A Jordanian Milk Pudding:
Muhallabyyah

Abstract The tradition of Middle East (ME) foods and beverages is linked with the history of traditional cuisine of several Asian areas. The diffusion of these traditions has determined the spreading of several ancient recipes in South Asia, the ME and in Europe also, with important influences concerfning the so-called 'Mediterranean Diet' (MD) lifestyle in particular.

Keywords Carbohydrate · Fat · Jordan · Mediterranean diet · Middle east · *Muhallabyyah* · Protein

Abbreviations

ITA   Italian Trade Agency
MD   Mediterranean Diet
ME   Middle East

6.1   Vegetables in the Middle East. Foods and Ingredients

The tradition of Middle East (ME) foods and beverages is linked with the history of traditional cuisine of several Asian areas. The diffusion of these traditions has determined the spreading of several ancient recipes in South Asia, the ME and in Europe also, with important influences concerning the so-called 'Mediterranean Diet' (MD) lifestyle in particular. The traditional cuisine in Countries such as Egypt, Iraq, Jordan, Lebanon, Saudi Arabia, Syria and Turkey is particularly variegated, from bread and bakery products to meat-based foods, from fish and seafood dishes to vegetable-based products (with the use of different fruits, nuts, rice, but also dairy products). This list concerns also the dessert products which have some important ingredients of vegetable origin, although the amount of needed ingredients does not rank in the first position. In this heterogeneous ambit, Jordan can exhibit some peculiarities.

© The Author(s), under exclusive license to Springer Nature Switzerland AG 2021   57
M. A. Haddad et al., *Chemical Profiles of Selected Jordanian Foods*,
Chemistry of Foods, https://doi.org/10.1007/978-3-030-79820-8_6

With relation to vegetable foods, the Jordanian Consumer is accustomed to use and prefer them as processed items, with or without important addition of non-vegetable ingredients. In this ambit, composite foods containing vegetables are important enough. A recent book has discussed these preferences in detail (Haddad et al. 2021), also linking consumeristic habits with healthy features of many fruits, vegetables and herbal preparations. The abundant amount of several active principles such as polyphenols (Haddad et al. 2020a, b–2021; Laganà et al. 2020) explains very well this aspect, although many other compounds may be investigated by the viewpoint of hygiene and safety, with relation to vegetable ingredients.

Four typical food preparations can show the importance of vegetable foods as ingredients, even if their role is only similar to decorations (El-Qudah 2015; Haddad et al. 2021; Khalifa and Shata 2018; Meneley 2007; Yamani and Mehyar 2011):

(a) *Halawa* or *Halva*. This product is a mixed dessert obtained from milk, sesame seed paste (*tahini*), cooked sugar, *Saponaria officinalis* (*halva* roots) and pistachios. Other ingredients and additives may be suggested. Anyway, the role of sesame seeds, pistachios and *halva* roots are the demonstration that this food needs a vegetable inclusion as one of its peculiarities. Interestingly, the *tahini* or *tahina* is a recurring ingredient in other typical ME dishes

(b) *Hummus*. This typical food, sometimes correlated with MD traditions, is obtained with different meaning, including dried chickpeas, garlic, the above-mentioned *tahini*, lemon juice and spices. It is widely diffused in the ME and non-Asian Countries such as the United States of America. This food can be consumed alone or as an appetiser

(c) The *muhallabyyah, muhallabieh, muhallabiyya,* or *muhallebi* speciality (Fig. 6.1). The basic recipe relies on (whole) milk and sugar. National differences can have some importance. Anyway, vegetable ingredients are at least rice (or corn) flour and rose water. One of the most interesting*muhallabyyah*versions are *Sahlab* or saloop, a hot-drink milk containing *Orchis anatolica* Boiss tubers (dried powder) as surrogate for corn or rice flours. Cinnamon powder should be also mentioned

(d) *Kunafeh, kunafa, knafeh,* or *kanafeh*. This product is a typical Arabic bakery product, obtained from flour, thick sugar syrup and goat's milk cheese (an 'extraneous' or non-vegetable ingredient).

As a result, the availability of ME and Jordanian foods with some vegetable peculiarity is assured. However, the variety of these preparations should take more than a single book. As a recent example, some Authors have discussed the matter of Sicilian street foods (SF), in particular when speaking of composite products containing rice (Barone and Pellerito 2020). The similarity in shape and basic design (including storage and transportation considerations) with several Arabic cuisine foods such as traditional *jameed* should be considered and thoroughly evaluated.

On these bases, the aim of this Chapter is to discuss the chemical profiles of one only of the above-mentioned foods: the *muhallabyyah* (Jessri et al. 2015; Saad and Ewida 2018).

Fig. 6.1 The *muhallabyyah, muhallabieh, muhallabiyya,* or *muhallebi* speciality. The basic recipe relies on whole milk and sugar. Vegetable ingredients are at least rice or corn flour and rose water Courtesy of Eng. Abdalla S. Alqawasmi (Bazaria Sweets Factory, Amman, Jordan)

6.2 Chemical Profiles of *Muhallabyyah*

On the basis of several available references on the Web, the chemical and nutritional profiles for *muhallabyyah/muhallebi* can be calculated (average data) as follows (per 100 g of food), as shown in Fig. 6.1:

- Energy (kJ): 1,615
- Energy (Kcal): 380
- Protein: 1.6 g
- Fat: 0.0 g
- Carbohydrates: 93.4 g (sugars: 70.7 g)
- Salt: 0.0 g.

Actually, some recipes might give very different results, with relation to fat content above all: the carbohydrate/fat/protein proportion could arrive to 3.0/ 1.1/ 1.0, as shown in Fig. 6.1 for a recipe containing animal butter (Anonymous 2021a; FatSecret 2021). Naturally, this result depends on the quantitative modification of main ingredients (milk, sugar, flour or starch) and on the possible introduction of animal fat (butter) (Siçramaz et al. 2016).

As observed for *shaneenah* (Chap. 2), the basic composition relies above all on three main ingredients. The remarkable amount of carbohydrates (>90%) is the function of milk, sugar and corn flour (or starch) at least; other minor ingredients may be allowed. Anyway, the ratio between milk, sugar and corn flour (or starch) is approximately 19.6/ 3.1/ 1 (with reference to flour assumed as '1'). The Turkish recipe for *kesme muhallebi* and also Egyptian recipes consider a preliminary mixture of these ingredients with starch, a subsequent pasteurisation (75 °C, 10 min) and the final addition of other ingredients, with subsequent mixing (5 min) and cooling at 4 °C (time: 24 h) (Khalifa and Shata 2018). The superficial addition of rose water before cooling is needed as a typical feature of the final pudding. Alternative recipes (Anonymous 2012; Anonymous 2021b; Russo 2017) may consider whole fresh milk, white sugar and rice flour in the ratio: 20/ 2/ 1 or 21.4/ 2.1/ 1 (additional ingredients: rose water and pistachios).

For these reasons, the *muhallabyyah* can be considered as an energetic product without fat-related dangers because of the chemical profile is completely towards carbohydrates. This situation can be different in certain versions such as *muhallebi* (Fig. 6.1) because of different quantitative indications and the introduction of animal fat: consequently, lipids and protein amounts can be higher than 4.0%, with a carbohydrate decrease until 33% (FatSecret 2021). On the other hand, some question could be made when speaking of microbiological resistance because of the notable amount of bioavailable carbohydrates… and the possible occurrence of aflatoxin production because of mould spreading over time (Ertas et al. 2011; Khalifa and Shata 2018) (Fig. 6.2).

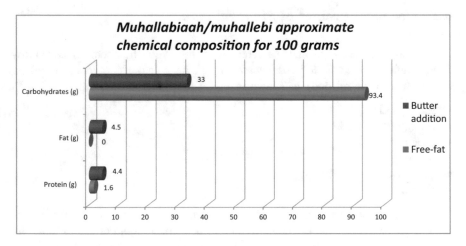

Fig. 6.2 The chemical and nutritional profiles for *muhallabyyah/muhallebi* demonstrate the absolute predominance of carbohydrates. The graph shows two possibilities, calculated for academic and research purposes only. Some recipes could give different results with relation to lipids and carbohydrates (Anonymous 2021a; FatSecret 2021). Naturally, this situation depends on the quantitative modification of main ingredients (milk, sugar, flour or starch) and on the possible introduction of animal fat (Siçramaz et al. 2016). Anyway, *muhallabyyah* can be considered as an energetic product without fat-related dangers because of the chemical profile is completely towards carbohydrates (between 33 and 93.4%)

References

Anonymous (2012) Muhallabia (dolce arabo di latte e farina di riso). Araba fenice in cucina! Available https://www.arabafeliceincucina.com/2012/05/golosi-anche-senza-glutine-muhallabia. html. Accessed 10 Feb 2021

Anonymous (2021a) Muhallabia dolce dal medio oriente senza glutine. blog.giallozafferano.it. Available https://blog.giallozafferano.it/timoelenticchie/muhallabia-dolce-dal-medio-oriente-senza-glutine/. Accessed 10th February 2021

Anonymous (2021b) Mohallabia/Muhallabia Recipe. RecipeOf Health.com. Available http:// recipeofhealth.com/recipe/mohallabia-muhallabia-297134rb. Accessed 10 Feb 2021

Barone M, Pellerito A (2020) Sicilian street foods and chemistry—the palermo case study. SpringerBriefs in Molecular Science. Springer, Cham. https://doi.org/10.1007/978-3-030-55736-2

El-Qudah JM (2015) Vitamin A contents per serving of eleven foods commonly consumed in Jordan. Int J Chem Tech Res 8(10):83–88

Ertas N, Gonulalan Z, Yildirim Y, Karadal F (2011) A survey of concentration of aflatoxin M1 in dairy products marketed in Turkey. Food Control 22(12):1956–1959. https://doi.org/10.1016/j. foodcont.2011.05.009

FatSecret (2021) Muhallebi. FatSecret. Available https://www.fatsecret.com/Diary.aspx?pa= fjrd&rid=3149415. Accessed 10 Feb 2021

Haddad MA, Dmour H, Al-Khazaleh JFM, Obeidat M, Al-Abbadi A, Al-Shadaideh AN, Al-mazra'awi MS, Shatnawi MA, Iommi C (2020) Herbs and medicinal plants in Jordan. J AOAC Int 103(4):925–929. https://doi.org/10.1093/jaocint/qsz026

Haddad MA, El-Qudah J, Abu-Romman S, Obeidat M, Iommi C, Jaradat DSM (2020) Phenolics in mediterranean and middle east important fruits. J AOAC Int 103(4):930–934. https://doi.org/ 10.1093/jaocint/qsz027

Haddad MA, Yamani MI, Jaradat DMM, Obeidat M, Abu-Romman SM, Parisi S (2021) Food traceability in Jordan. Current perspectives. SpringerBriefs in Molecular Science. Springer, Cham

Khalifa M, Shata RR (2018) Mycobiota and aflatoxins B1 and M1 levels in commercial and homemade dairy desserts in Aswan City. Egypt. J Adv Vet Res 8(3):43–48

Jessri M, Farmer AP, Olson K (2015) A focused ethnographic assessment of middle eastern mothers' infant feeding practices in Canada. Matern Child Nutr 11(4):673–686. https://doi.org/ 10.1111/mcn.12048

Laganà P, Coniglio MA, Fiorino M, Delgado AM, Chammen N, Issaoui M, Gambuzza ME, Iommi C, Soraci L, Haddad MA, Delia S (2020) Phenolic substances in foods and anticarcinogenic properties: a public health perspective. J AOAC Int 103(4):935–939. https:// doi.org/10.1093/jaocint/qsz028

Meneley A (2007) Like an extra virgin. Am Anthropol 109(4):678–687

Russo R (2017) "La porti un..." cedro del Libano: Muhallabia! Il sorriso vien mangiando. Available http://www.ilsorrisovienmangiando.com/2017/07/la-porti-un-cedro-del-libano-muhallabia.html. Accessed 08 Feb 2021

Saad NM, Ewida RM (2018) Incidence of cronobacter sakazakii in dairy-based desserts. J Adv Vet Res 8(2):16–18

Siçramaz H, Ayar A, Ayar EN (2016) The evaluation of some die-tary fiber rich by-products in ice creams made from the traditional pudding–kesme muhallebi. J Food Technol 3(2):105–109. https://doi.org/10.18488/journal.58/2016.3.2/58.2.105.109

Yamani MI, Mehyar GF (2011) Effect of chemical preservatives on the shelf life of hummus during different storage temperatures. Jordan J Agric Sci 7(1):19–31

Printed in the United States
by Baker & Taylor Publisher Services